Lecture Notes in Mobility

The book series Lecture Notes in Mobility (LNMOB) reports on innovative, peer-reviewed research and developments in intelligent, connected and sustainable transportation systems of the future. It covers technological advances, research, developments and applications, as well as business models, management systems and policy implementation relating to: zero-emission, electric and energy-efficient vehicles; alternative and optimized powertrains; vehicle automation and cooperation; clean, user-centric and on-demand transport systems; shared mobility services and intermodal hubs; energy, data and communication infrastructure for transportation; and micromobility and soft urban modes, among other topics. The series gives a special emphasis to sustainable, seamless and inclusive transformation strategies and covers both traditional and any new transportation modes for passengers and goods. Cutting-edge findings from public research funding programs in Europe, America and Asia do represent an important source of content for this series. PhD thesis of exceptional value may also be considered for publication. Supervised by a scientific advisory board of world-leading scholars and professionals, the Lecture Notes in Mobility are intended to offer an authoritative and comprehensive source of information on the latest transportation technology and mobility trends to an audience of researchers, practitioners, policymakers, and advanced-level students, and a multidisciplinary platform fostering the exchange of ideas and collaboration between the different groups.

More information about this series at https://link.springer.com/bookseries/11573

Gereon Meyer · Sven Beiker
Editors

Road Vehicle Automation 9

Springer

Editors
Gereon Meyer
VDI/VDE Innovation + Technik GmbH
Berlin, Germany

Sven Beiker
Stanford University
Palo Alto, CA, USA

ISSN 2196-5544 ISSN 2196-5552 (electronic)
Lecture Notes in Mobility
ISBN 978-3-031-11111-2 ISBN 978-3-031-11112-9 (eBook)
https://doi.org/10.1007/978-3-031-11112-9

This Springer imprint is published by the registered company Springer Nature Switzerland AG
The registered company address is: Gewerbestrasse 11, 6330 Cham, Switzerland

Preface

After more than a decade of research and development since famous Silicon Valley companies got engaged in automated road transportation, the field seems at a crossroads: The technology options are on the table. And, numerous demonstrations and pilot projects have been running all around the globe, showcasing the opportunities of automated mobility for a multitude of applications: highway trucks, urban goods delivery, shared passenger shuttles, and individual robotaxi and for extending the range of public transport over the last mile.

Yet, to turn these prospects into marketable products and service offerings, some fundamental technical choices have to be made, e.g. whether the decision-making capabilities will be located at the edge of a car's electronic control systems, or whether these require some kind of greater intelligence in the cloud, or maybe a mixture of both. Which way to choose is less a question of right or wrong, but merely of the desired level, coverage and reliability of the automated function, as expressed in the Operational Design Domain. And, it will have tremendous impact on the requirements for the vehicle's electronic and software control architecture, the bandwidth and latency of the data communication with the digital infrastructure and the quality of the decision-making process. Hence, the lessons learned from pilots and demonstrations on societal needs, market relevance, legal feasibility, user expectations, and yes, also technology maturity are essential to co-create the relevant technology paths.

Not least in view of the disruptions of worldwide supply chains, most notably for raw materials sourcing and semiconductor chip making, it is of high strategic importance for the technological sovereignty to keep and reach these decisions wisely, timely, and in a coordinated way. Otherwise, automated road transportation may even risk to fail due to inconsistencies between projected opportunities and imminent challenges, or be available only at the cost of painful compromises in terms of the democratic values of safety, equitable access, and data privacy. The comprehensive and worldwide dialogue among all players of the automated road transportation ecosystem including technologists, planners, operators, legislators, and last but not least citizens, therefore, is as important as never before.

That said, we are proud that the expert discussions on automated road transportation have outlived the challenges of the COVID-19 pandemic and are again well reflected in this ninth edition of the Road Vehicle Automation book series. The chapters of this volume have been written by individuals or teams that had presented and discussed their latest findings at the inaugural Automated Road Transportation Symposium (ARTS), firstly launched by the Transportation Research Board (TRB) in succession of the Automated Vehicles Symposium (AVS) as a virtual event in July 2021.

While we are looking forward to reconnect with the authors and the automated mobility community at ARTS22, and planning for the next volume, we would like to express our gratitude to all contributors for the time and efforts they spent on their chapters, to Jane Lappin, Valerie Shuman, and Steven Shladover from TRB for their continuous support, and to colleagues at VDI/VDE-IT, particularly the working students of the European and International Business Development Department, Laura Soto, Jacques Dalhoff, and Paul Mengeling, for assistance in the editorial process. Moreover, we would like to express our gratitude to Springer teams in Germany and India for making this book publication possible in a fast, reliable, and high-quality manner as usual.

May 2022

Gereon Meyer
Sven Beiker

Contents

Introduction: The Automated Road Transportation Symposium 2021 1
Steven E. Shladover, Jane Lappin, and Valerie Shuman

Part I: Public Sector and Policy Activities

Development of Driving Intelligence Validation Platform (DIVP®) for ADS Safety Assurance 13
Seigo Kuzumaki and Hideo Inoue

The Great Promise of AV - But it Needs to be Done Properly 23
Robert L. Sumwalt

How Critical is Connectivity? 34
Katherine Kortum

Innovation Strategies and Funding Policies for Automated and Electric Road Mobility 42
Gereon Meyer, Carolin Zachäus, and Jakob Michelmann

Part II: Business Models and Operations

Automated Vehicle Fuel Efficiency Town Hall 53
Dimitris Assanis, Avi Chaim Mersky, Jihun Han, Therese Langer, Jeffrey Lidicker, Rick Mihelic, Gabor Orosz, and Marina Sofos

Sharing Automated Vehicles: Policies and Ideas to Improve the Sharing Experience to Reduce Congestion and Energy Use in a Post-COVID World 71
Joshua Auld, Denise Baker, Danielle Chou, Ann Foss, Elizabeth Machek, William Riggs, and Scott Smith

Part III: Users and Human Factors

Human-Centric Intelligent Driving: Collaborating with the Driver to Improve Safety 85
Avinash Balachandran, Tiffany L. Chen, Jonathan Y. M. Goh, Stephen McGill, Guy Rosman, Simon Stent, and John J. Leonard

Automated Shuttles and Buses for All Users 110
Katherine F. Turnbull

Part IV: Vehicle and Road Systems Technology Development

What's Next in AV Standards and Simulation Validation? 123
Shawn Kimmel, George Nicols, Edward Straub, and Edward Chow

Six Years of Reading the Road Ahead: Supporting Roadway Automation with Traffic Control Devices 140
Scott O. Kuznicki

Remote Support for Automated Vehicle Operations 149
Jaap Vreeswijk, Azra Habibovic, Olav Madland, and Floris Hooft

Part V: Transport System Planning

Ensuring Strong Public Support for Automation in the Planning Process: From Engagement to Co-creation 167
Leah Kaplan, Siegfried Rupprecht, Monica Grosso, Nikolas Thomopoulos, Wolfgang Backhaus, Maria Alonso Raposo, Biagio Ciuffo, and Daniel Franco

Author Index 185

Introduction: The Automated Road Transportation Symposium 2021

Steven E. Shladover[1](✉), Jane Lappin[2], and Valerie Shuman[3]

[1] University of California PATH Program, 1357 South 46th Street, Building 452, Richmond, CA 94804, USA
steve@path.berkeley.edu
[2] TRB Vehicle-Highway Automation Committee, Belmont, MA, USA
[3] Shuman Consulting Group, LLC, TRB Challenges and Opportunities of Road Vehicle Automation Subcommittee, Skokie, IL, USA
vs@shumangroupllc.com

Abstract. In 2021, the Automated Vehicles Symposium was succeeded by the Automated Road Transportation Symposium (ARTS21), which was produced entirely by the Transportation Research Board. With the continuing disruptions from the global pandemic, it was again produced as a virtual online meeting. The plenary presentations and breakout discussions continued to provide the meeting participants with the most up-to-date and authoritative information about the current international state of development and deployment of road vehicle automation systems, retaining its standing as the essential global meeting for industry, government and research practitioners in the field.

Keywords: Road vehicle automation · Road transport automation · Automated vehicles · Autonomous vehicles · Self-driving vehicles

1 Overview

The 2021 Automated Road Transportation Symposium was organized and produced by the National Academies of Science, Engineering and Medicine (NASEM) Transportation Research Board (TRB), following the withdrawal of the Association for Unmanned Vehicle Systems International (AUVSI). This meeting was organized to serve the participants' interests in understanding the impacts, benefits, challenges and risks associated with increasingly automated road vehicles and the environments in which they operate. It brought together key government, industry and academic experts from around the world with the goal of identifying opportunities and challenges and advancing Automated Driving System (ADS) research across a range of disciplines.

The symposium took place online over four days, 12–15 July, 2021, with session times scheduled to accommodate participants in Europe and East Asia as well as North America. The plenary sessions were scheduled from 9 am – 12 noon on the first day, 12 noon to 2 pm on the second and third days, and 1 to 4 pm on the last day (all Eastern Daylight Time), to provide diverse opportunities for participants outside North America

G. Meyer and S. Beiker (Eds.): ARTSymposium 2021, LNMOB, pp. 1–10, 2023.
https://doi.org/10.1007/978-3-031-11112-9_1

to participate. and all were recorded to be available for later viewing by registrants who could not participate at the scheduled times. Audience questions were relayed through the online chat function to augment the questions that plenary panel moderators prepared for the discussions with their panelists.

Breakout sessions were scheduled in three-hour blocks on the first afternoon and the mornings of the other three days plus two-hour blocks on the two middle days. Each time slot was able to accommodate up to six parallel breakout sessions, based on the constraints imposed by the online platform. The mixture of morning and afternoon breakout session times on the US east coast provided opportunities for participation by European and Asian attendees without too much difficulty in some of the breakouts.

The breakout sessions were organized by committees of volunteers to address a wide range of topics. These were clustered into four thematic tracks to make it easier for attendees to identify the sessions of strongest interest to them:

- Policy and Planning
- Users and Human Factors
- Operations and Applications
- Technology.

The plenary and breakout session programs were planned and produced by the ARTS21 Planning Committee, which included a mixture of TRB volunteers and support staff from the US DOT Volpe Center:

Molly Behan, Volpe Center; Richard Bishop, Bishop Consulting; Richard Cunard, Engineer of Traffic and Operations, TRB; Kevin Dopart, U.S. DOT Intelligent Transportation Systems Joint Program Office, Cynthia Jones, Drive Ohio; Jane Lappin, TRB Vehicle-Highway Automation Committee Chair; Jarred Myers, Volpe Center; Steven Shladover, University of California PATH Program (and former chair of the TRB Vehicle-Highway Automation Committee); Valerie Shuman, Shuman Consulting Group, LLC and Chair, TRB CORVA Subcommittee; and Edward Straub, SAE.

2 Keynote Talks

The symposium began with a keynote talk by the recently-retired Chairman of the National Transportation Safety Board, Robert Sumwalt.

Mr. Sumwalt spoke to the potential of automated vehicles to significantly reduce fatalities and injuries caused by roadway crashes, and to eliminate roadway congestion. The devil is in the details, he said, and to achieve these benefits, the development and implementation must be done properly. In his talk, and in the chapter he wrote from his talk for this edition (Sumwalt 2023), he offers his suggestions to manufacturers and regulatory authorities for how to minimize unintended consequences of highly automated vehicles and ensure proper introduction of AV technology.

U.S. Senator Gary Peters from Michigan, who has been an advocate for legislation covering road vehicle automation issues, gave a brief pre-recorded keynote talk addressing the importance of establishing a strong federal framework for automated driving in the U.S., emphasizing the need to invest resources to compete effectively with China and create a transformative future. He spoke to the potential of connected and automated vehicle technologies to save lives, open new mobility options for the elderly and those with intellectual and physical impediments to driving, and he stressed that we must ensure that automated vehicles research and production is done in the USA. Senator Peters closed his keynote by reiterating his commitment to safe testing and deployment of automated vehicles.

The third keynote event was a discussion between Dr. Chris Urmson, the CEO of Aurora Innovations and Dr. Steven Cliff, the Deputy Administrator of NHTSA, moderated by Marjory Blumenthal of RAND. They discussed the complexity of encoding traffic interactions, safety culture, the convergence of the trends toward electrification and automation of road transportation, and the advantages of using professionally managed fleets of vehicles as the initial platforms for ADS deployment.

Dr. Urmson and Dr. Cliff agreed on the challenge of encoding driving complexity; Dr. Cliff illustrated this with a reflection on his experience as a motorcyclist to describe the subtle signals exchanged among drivers and other road users that indicate intent. Dr. Urmson described the likely safety impact of automated driving systems over time: he anticipates that the first Automated Driving Systems (ADS) will adapt to the local driving habits, and over time, as there are more ADS on the road, local driving culture may adapt to the safe driving profiles of the ADS. Discussing safety culture, he offered examples from Aurora of their safety practices covering the full vehicle life cycle and the development and implementation of a "Safety Management System." Dr. Cliff remarked that safety culture, like equity and access, should be built into the vehicle system from the start. He stressed that policies need to drive the outcomes we want to achieve, citing the importance of improved transportation access, equity, affordability, and zero emissions.

Ms. Blumenthal asked about the convergence of electrification and automation, and about the impact of new ADS business models on public policy objectives. Dr. Urmson agreed that while automation and electrification will converge in the future, Aurora must focus today on achieving automation safely, quickly, and broadly if they are to achieve their company and policy goals. He said that the fleet operation business model, versus private ADS ownership, would enable Aurora to accomplish their objectives more quickly, to manage the fleet more safely, and provide better access to transportation for all. Dr. Cliff strongly stated the pressing importance of achieving zero vehicle emissions. He agreed that the ADS fleet model promised greater transportation safety, citing the success of the airline safety regulatory framework. In closing the discussion, Dr. Cliff reiterated NHTSA's objectives to ensure that ADS achieve their promise of improved traffic safety for all road users, reduced tailpipe emissions, and increased mobility for people, goods, and services.

Dr. Steven Cliff also provided the keynote address on the final day of the symposium. He spoke about the importance of a safe systems approach to ensure road safety for all users, building from the five road safety E's: equity, engineering, education, enforcement, and emergency medical service. Dr. Cliff said that the development and deployment of safe, equitable, clean, accessible, and mature automated driving systems is a top commitment from the USDOT and NHTSA, noting that cautious and responsible progress is the right tempo for development. He addressed critical ADS policy issues, such as job impacts, privacy, and access, emphasizing that equity must be planned into the future of ADS to achieve a safe and equitable future for all.

Dr. Cliff spoke about current NHTSA actions and ADS research activities. He reported a recent NHTSA action requiring that manufacturers report all crashes for level 2 ADAS-equipped vehicles and prototype ADS vehicles, with one-day reporting deadlines for the most serious crashes. NHTSA is doing research on ADS test tools and methods, crash behavior, overall system behavior, critical subsystem performance, validation and verification, and human factors, including external interaction with other vehicles and vulnerable road users. In closing, he said that everything that NHTSA does is focused on ensuring the safety of American people on our roads no matter how they travel.

Meera Joshi, the Deputy Administrator of the Federal Motor Carrier Safety Administration (FMCSA), responsible for the safety of operations of interstate buses and commercial trucks, spoke about achieving increased safety for commercial vehicle drivers and all road users through testing and deployment of advanced driver safety systems and automated vehicles. She reported that FMCSA is working closely with NHTSA on a regulatory framework to achieve the right balance between automated vehicle innovation and accountability, and to address complex issues, like automated emergency braking and roadside safety inspections. While automation can increase future work opportunities for today's drivers, she noted that industry and government will need to make a significant investment in advanced safety technologies and driver training to ensure safe driving and successful jobs transitions. In closing, she highlighted the importance of research to inform policy and, specifically, the importance of TRB research to the work of FMCSA.

3 Plenary Panel Sessions

ARTS21 extended the trend from previous years of devoting a majority of the plenary program time to panel discussion sessions on important topics, featuring groups of speakers responding to questions from the moderator and interacting with each other, to break up the sequence of formal presentations. These also provided opportunities for audience members to submit questions through the online chat function. Three of these panel sessions focused on applications of automation to specific transportation services, and four sessions were devoted to more general cross-cutting topic areas.

3.1 Opportunities in Automated Local Package Delivery

One of the automation applications that has seen the most rapid growth in activity in the last few years is delivery of small packages over short distances in urban and suburban environments. Steven Shladover moderated a discussion with representatives of Starship and Refraction AI, two companies that have developed small special-purpose vehicles for urban delivery of packages. They discussed the technical simplifications that are possible when the vehicles do not need to accommodate human passengers, and how they decided to focus on the special vehicle designs that they each chose for operating on sidewalks and in bicycle lanes rather than mixing with full-scale vehicle traffic.

3.2 Automated Trucking

The other major goods movement application for automation is at the opposite end of the spectrum, which is moving goods in large long-haul trucks and mid-size trucks from distribution centers to retailers. Richard Bishop moderated this panel discussion with representatives of Gatik and Embark Trucks and their customers Loblaw and Hewlett-Packard, as well as the U.S. Army. These organizations are using automation for high-speed highway driving rather than in higher density urban areas, and expect to see the drivers who are no longer needed for these driving assignments shifting toward local package delivery driving, where they would still have responsibilities for taking the packages the final 50 ft. to the customers.

3.3 Automation in Shared Mobility and Public Transit

The passenger-carrying applications with the highest level of activity have been in shared mobility and public transit, which were explored in this session moderated by Kelley Coyner, with representatives from New Flyer, Connecticut DOT and the Jacksonville Transit Authority. They discussed fixed-route transit applications of automation on dedicated guideways and partial driving automation, which simplify the technological challenges and make it possible to consider nearer-term deployments.

3.4 Understanding Critical Challenges for Safer Deployment of AVs with the Automated Vehicle Safety Consortium (AVSC)

The Automated Vehicle Safety Consortium (AVSC) was established as one of SAE's Industry Technologies Consortia (ITC) to provide a mechanism for companies to cooperate on pre-competitive aspects of AV safety. Amy Chu, the Director of AVSC, moderated a panel discussion with representatives from member companies Lyft, Aurora, Ford and Toyota. They discussed a wide variety of important issues associated with developing safety cases, minimizing safety risks throughout the development process, and establishing meaningful safety measures of effectiveness and bases for comparison with human driving safety. This provided a good indication of the serious industry efforts to grapple with the safety challenges of automation and the need to build public trust in ADS safety.

3.5 What's Ahead for AV Legislation and Regulations?

The United States has not yet passed any legislation or created any regulations to address the many issues associated with public deployment of automated driving systems. This has been in large part because of fundamental disagreements between the industrial organizations developing the technology on one hand and the traffic safety advocates and government agencies responsible for transportation safety on the other hand. These challenges were evident in the discussion in this panel session moderated by Sam Mintz of Politico, with representatives from the Alliance for Automotive Innovation, the Consumer Federation of America, Consumer Reports, The American Association of Motor Vehicle Administrators and Venable LLP. There were differences of perspective regarding approaches to regulation, the *a priori* assumptions that should be made about safety of automated driving, and the relative priorities to assign to driving assistance systems versus automated driving systems, with little indications that some middle ground could be found as a basis for establishing clear government policy.

3.6 State DOT Automated Vehicle Research and Collaboration Activities

Nick Hegemier from the Drive Ohio program moderated a discussion with representatives from the Minnesota, Texas and Maryland DOTs regarding their respective activities.

3.7 The Business of Automated Vehicles and the Path to Commercialization

Grayson Brulte of Brulte & Company led a lively discussion exploring the state of the industry with Alan Ohnsman from Forbes, Joann Muller from Axios Navigate, and David Welch from Bloomberg News. They brought the Symposium to a stimulating close with insightful observations about the current challenges that the AV industry faces. They discussed the day's main news item about Aurora going public through a SPAC, and then emphasized the challenges the entire industry is facing in identifying viable business cases that can earn profits within the relatively near term (leading to the current emphasis on trucking applications). None of them envisioned large-scale public deployments occurring within the next three or four years, but they expected to see extensive consolidation of the companies being necessary for survival.

4 Plenary Presentations

Individual presentations were distributed across the plenary program in between the panel discussions to avoid Zoom fatigue from too long a sequence of consecutive presentations. Six of the presentations were given by speakers who were invited to cover specific topics that the planning committee believed to be important for the audience to learn about and the other seven presentations were progress reports on some of the most important public-sector activities around the world related to automation (three within the U.S. DOT and the other four international).

4.1 Presentations on Specific Topics

- Mykel Kochenderfer and Steve Moss, Stanford University - Validation of Safety-Critical Decision Making Systems
- Pnina Gershon, MIT - Driver Behavior and the Use of Automation in Real-World Driving
- Kristin Kolodge, J.D. Power - Voice of the Customer: Experience Matters
- Avinash Balachandran, Toyota Research Institute - Guardian: Sharing Control with the Driver to Improve Safety
- Mauricio Peña, Waymo - Safety at Waymo
- Alexander Kraus, IAMTS—IAMTS -- An Alliance to Internationally Accelerate and Harmonize Safe Deployment of Connected and Automated Vehicles

4.2 National and International Government Activities Relevant to Automated Driving

- Kevin Dopart, USDOT ITS Joint Program Office - U.S. DOT Automated Vehicle Research Activities
- Jane Doherty, NHTSA - Why the UNECE Matters to the Future of Automated Vehicles
- Tom Alkim, European Commission – Connected, Cooperative and Automated Mobility in Europe
- Jenny Laber, UK Department for Transport – Shaping the Future of Transport: Automated Mobility in the UK
- Seigo Kuzumaki, SIP-adus Program (Japan) – Development of "Driving Intelligence Validation Platform" for ADS Safety Assurance

5 Breakout Sessions

ARTS breakouts gather key experts from around the globe for more in-depth consideration of specific topic areas. The goal of the breakout sessions is to collaboratively answer the questions: *What needs to be true to make the AV vision become a reality? How can our research help drive progress year on year?* The 2021 program included 31 sessions with nearly 450 speakers and covered a wide range of specialized topics from across the field to enable this discussion for the industry as a whole (see program list below).

The primary findings from each afternoon's breakout discussions were reported back to the plenary the following morning. The combined summaries provided in these Daily Roundups distill the latest insights from across the industry, including:

- *Collaboration areas* are coming into focus. The 2020 Symposium conversations consistently called for cross-domain, cross-sector, cross-industry, cross-jurisdiction and/or global collaboration. One year later, the 2021 conversation highlighted industry convergence on a consistent set of immediate focus areas for this work, which are currently being articulated as: User Needs, Standards and Metrics, particularly as they apply to Regulatory, Operations and Technology areas.
- Extending both *ODDs and accessibility* were additional areas which are receiving major global emphasis. There is intense interest in understanding, defining and extending the ODD coverage of automated systems. On the accessibility side, the Symposium updates and discussions reinforced the industry goal of using automated solutions to expand the suite of mobility options across both ability and economic spectra.
- The industry *knowledge base* is expanding. Many topic areas reported that they have now amassed enough experience to start building shareable libraries of insights and best practices. Similarly, the research and testing underway has started to yield consistently positive user feedback, providing helpful momentum for further efforts.
- There is an increasing focus on the practical details of *sharing* a broad range of data across disciplines and sectors. Digital twins are seen as a promising tool to help this process.
- Initial *transit, delivery and freight* automation are making real progress. The intense focus on logistics driven by the pandemic has fueled a major increase in efforts to automate the entire supply chain, and there were many reports from pilots and initial deployments which have successfully delivered measurable benefits.

5.1 ARTS Breakout Sessions

5.1.1 Policy and Planning Sessions

- An Inside Look at Policy Making for Automated Vehicles
- Efficiency Town Hall: AV Fuel Economy & Efficiency Regulations and Technologies
- Shark Tank: Everything From Free Freight to AV for Low Income Travelers to How Many AV Firms Will Survive
- Ensuring Strong Public Support for Automation in the Planning Process
- Sharing AVs: Policies and Impacts in a Post-COVID World
- New Horizons for Connected, Cooperative & Automated Mobility (CCAM) in Europe
- Proxy Metrics for Social Equity Considerations of Automated Vehicles
- The Long and Winding Road: Planning and Network Analysis for CAVs
- Ticket to Ride: City Council Hearing on Autonomous Transit Vehicle

5.1.2 Users and Human Factors Sessions

- Inclusive by Design: Creating an Equitable and Accessible Automated Future
- AV to Road User Communications: What Have We Learned from Research?
- Older Adults & Automated Driving Systems

5.1.3 Operations and Applications Sessions

- Trucking Automation: Delivering Freight on Automated Trucks Today
- Driving AV Data Exchange between Public and Private Sectors
- AVs in Rural America - Equity and Mobility for All
- Designed, Wheeled, Delivered II: Scaling Up Automated Urban Delivery Vehicles and Devices
- Remote Support for Automated Vehicle Operations
- The Managed and Shared Roads to AV Deployment: How can we get there?
- Advances in Automated Transit Buses
- Integrated Traffic Management and CVs/CAVs for Freeways and Arterials
- Public and Private Sector Collaboration to Advance Automated Driving Systems Testing and Deployment
- Automated Shuttles and Buses for All Users
- Reading the Road Ahead: Highway Agency Efforts in Supporting ADAS & HAV Integration with the Roadway Environment

5.1.4 Technology Sessions

- Safety Assurance of Automated Driving
- Automated Driving System (ADS) Safety Metrics in Theory and Practice
- Computational & Algorithmic Challenges for AI Applications in the Era of CAVs
- Enabling Technologies - A Peek Under the Hood
- What's Next in AV Standards?
- ADS Simulation and Testing Part 1: What's New?
- ADS Simulation and Testing Part 2: Approaches for Collaboration & Validation
- Preparing for AVs and Shared Mobility: How Critical is Connectivity?

6 General Cross-Cutting Observations

As the field of road vehicle automation has advanced and the level of knowledge of the issues has grown over the past several years, the areas of emphasis within the Symposium have continued to evolve. Based on the discussions at this most recent meeting, several general observations are worth noting:

- Although the COVID-19 pandemic has caused widespread disruptions throughout the world, it does not appear to have produced fundamental changes to the development of driving automation systems. It has, however, accelerated several trends that were already evident in the industry, including the consolidation and partnering of companies developing full-stack automation systems and the shift of emphasis towards goods movement in preference to passenger movement.
- The goods movement applications already had several factors working in their favor (especially strong business cases and reduced technical complexity compared to passenger movement), but these were amplified by pandemic conditions, with the dramatic growth in interest in home delivery of restaurant meals, food and a wide range of retail goods.
- The automated ride hailing applications (which were previously the "star" attractions for automation) lost some of their luster based on the growing recognition of the severe technological challenges they face in delivering a quality of service that will be acceptable to passengers, combined with pandemic-related discomfort with sharing a confined space with strangers of unknown health status.
- The pandemic disruptions to working conditions, especially for activities that require close personal interactions such as testing and debugging vehicle technologies, slowed the pace of development and testing work sufficiently to push more of the automation developers to consolidate with new partners in order to survive the extended development cycle.
- There has also been a growing consensus that the applications of higher automation for the foreseeable future will be on commercially operated fleet vehicles rather than private personal vehicles. This is primarily an economic consideration, because the vehicle technology will be so expensive that it will only be affordable in commercial operations that can reduce their operating costs significantly (eliminating driver labor costs) and can productively operate the vehicles for many hours per day to amortize the investment.
- The automation system developers are challenged to find the intersection between transportation applications that have a large enough market to be commercially viable and the applications that are technologically feasible. This can best be translated into identifying the viable operational design domain (ODD) for each system. Tightly constrained ODDs are much closer to being feasible than unconstrained ODDs, which tends to favor limited locations with low densities of vehicle and VRU traffic, low speeds and mild weather. On the other hand, locations with higher traffic densities, higher speeds, and more variable weather offer more attractive commercial opportunities.
- The concerns about potential losses of driver jobs that were discussed in previous years appear to have receded this year based on increased recognition of the severity

of the driver shortage and of how gradual the rollout of automated driving will be. This has also been mitigated by growing understanding of the importance of remote human support for enabling automated driving systems to cope with the full complexity of the driving environment. Drivers who are no longer employed in the driver's seat of a commercial truck or bus or taxi could still find employment opportunities in the fleet management operations centers, where they would be monitoring the operations of driverless vehicles and intervening to help them manage situations that their ADS do not understand.

- The importance of developing and effectively communicating safety cases for automated driving systems was highlighted. The safety cases are needed to provide sufficient evidence of safety prior to widespread deployment of systems, but they must be presented in "explainable" form to serve the purpose of earning trust from safety regulators, traffic safety stakeholders and the general public. Mechanisms are needed for facilitating effective communication between the system developers and the other stakeholders throughout the development process.
- There has not been much progress on developing a legislative and regulatory framework for the deployment of automated driving in the U.S. because there are large differences in perspectives and priorities between the system developers on the one hand and the consumers, safety advocates, and transportation agencies on the other hand.
- The complexity of the intended ODD for each automated driving system was recognized to be the dominant factor in determining the timing of deployment (and the relative timing between different automated applications).

Reference

Sumwalt, R.L.: The great promise of AV - but it needs to be done properly. In: Meyer, G., Beiker, S.(eds.) ARTSymposium 2021, LNMOB, pp. xx–yy. Springer, Cham (2023). https://doi.org/10.1007/978-3-031-11112-9_3

Part I: Public Sector and Policy Activities

Development of Driving Intelligence Validation Platform (DIVP®) for ADS Safety Assurance

Seigo Kuzumaki[1(✉)] and Hideo Inoue[2]

[1] Toyota Motor Corporation, Toyota, Japan
seigo_kuzumaki@mail.toyota.co.jp
[2] Kanagawa Institute of Technology, Atsugi, Japan
inoue@cco.kanagawa-it.ac.jp

Abstract. SIP, or Cross-ministerial Strategic Innovation Promotion Program is a 5 year R&D program led by the Japan government. SIP-adus, or automated driving system for universal service aims to realize automated driving systems (ADS) with Government-Industry-Academia cooperation. SIP-adus focuses on cooperative R&D themes such as Dynamic Map, Safety assurance, Cybersecurity, and so on. Safety assurance is quite important for spread of ADS. A collaborative consortium was established for the development of Driving Intelligence Validation Platform (DIVP®) [1] for Automated Driving Safety Assurance. This research project is developing an assessment platform using simulators to create a virtual model by adopting a series of models consisting of driving environments, spatial propagation, and sensors that are highly consistent with the actual phenomena required to assess the safety of automated driving. The goal and current status of DIVP® activity is explained in this report.

Keywords: Automated driving · Safety assurance · Simulation · Sensor model · Interface

1 Introduction

Safety is critical and essential for ADS realization. In order to realize and spread ADS in the market, it is necessary to have a feasible methodology to assess the safety performance of the vehicle.

ADS conduct "recognition", "judgement" and "operation" continuously on behalf of the human driver. For this reason, sensor performance for recognition is a significant factor for ADS safety. SIP is working to develop a platform to evaluate the safety of ADS in a virtual space. The project is named DIVP® – Driving Intelligence Validation Platform. This project focuses on a precise duplication from real to virtual, and on sensor models' verification of consistency with real word testing. DIVP® objectives are to define open standard interfaces, to establish a 'reference platform' with a reasonable verification level, (especially, for sensor modeling), and to establish the Environment & Sensor pair model-based approach for Validation & Verification reality. (Fig. 1).

G. Meyer and S. Beiker (Eds.): ARTSymposium 2021, LNMOB, pp. 13–22, 2023.
https://doi.org/10.1007/978-3-031-11112-9_2

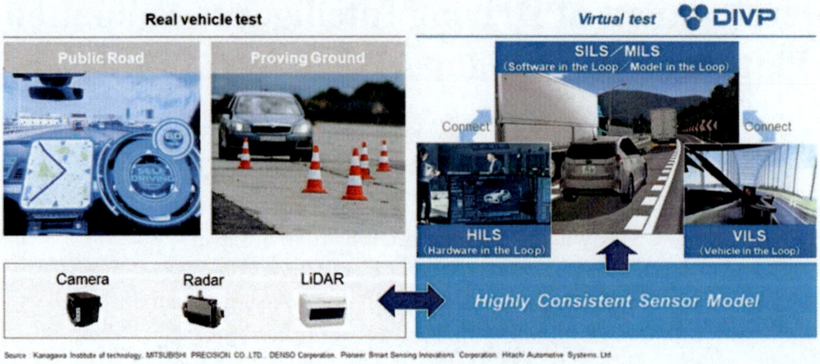

Fig.1. Overview of DIVP® project

2 Driving Intelligence Validation Platform (DIVP®)

SIP-adus launched the DIVP® research and development project as a collaborative consortium between sensor manufacturers, software companies, universities, and other parties, with the aim of building a virtual automated driving safety assurance simulation platform at the end of 2018. (Fig. 2) In general simulators, the main focus is on evaluating whether the system control works correctly, and many sensing models are based on so-called ground truth models, that is to say a functional model.

But in order to guarantee the safety of an automated driving vehicle, it is very important to verify the physical limitation of sensors as environmental monitoring functionality, such as cameras, radars, and LiDARs. So, our goal is to establish a simulation platform to be able to validate the safety assurance including sensor performance.

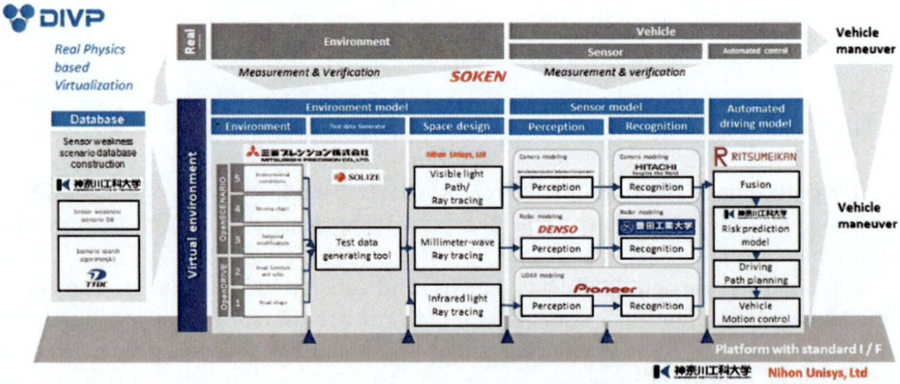

Fig. 2. Configuration of DIVP® project

2.1 Scope and Objective

We develop a spatial propagation model of the ray tracing system based on the reflection characteristics and transmission characteristics. DIVP® scope covers especially,

"Physical Model improvement", "Computing Performance" and "Data accumulation & utilization" in Trinitarian approach.

Based on this approach, DIVP® objectives are to define open standard interfaces, to establish a 'reference platform' with a reasonable verification level, especially, for sensor modeling, and to establish the Environment & Sensor pair model-based approach for Validation & Verification reality. (Fig. 3).

Our purpose is a contribution to ADS safety assurance as a methodology platform, so we think that international cooperation for common use is very important.

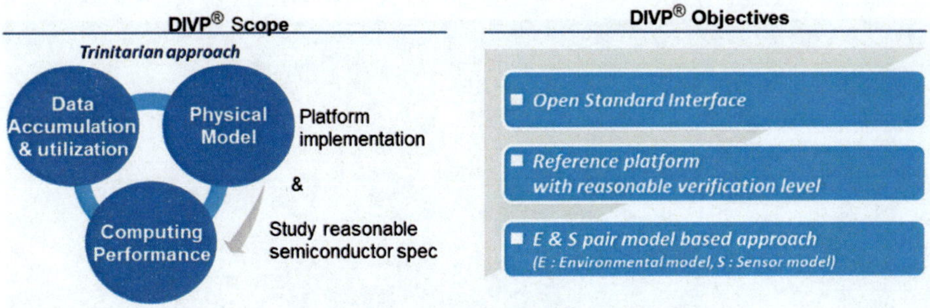

Fig. 3. Scope and objective

2.2 Sensor Modeling Based on Physical Property Measurement

This project is building physical models of the reflection characteristics (including the reflexive characteristics, diffusion, and specular reflection) and transmission characteristics of millimeter wave radar waves, visible light for cameras, and near-infrared light from LiDAR, constructing them as spatial propagation models of ray tracing and the like. At the same time, the project is also working on creating physical models of physical phenomena affected by the surrounding environment, such as rain, fog, sunlight, and so on. Specific examples of each of these models are described below. (Fig. 4).

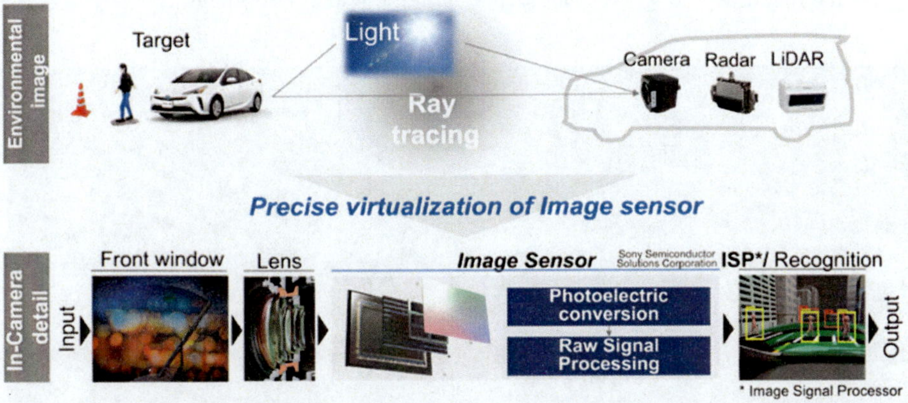

Fig. 4. Sensor model to simulate spatial propagation (for Camera)

2.2.1 Camera Model

The DIVP® camera model simulates spectral characteristics that are input to semiconductors such as CMOS, rather than the RGB is human eyes friendly.

In addition, sunlight is formularized as a sky model, which allows the precise simulation of solar light sources based on time, latitude, and longitude inputs. As described below, a realistic sensor simulation view has been realized by defining the reflection characteristics of objects as property. (Fig. 5).

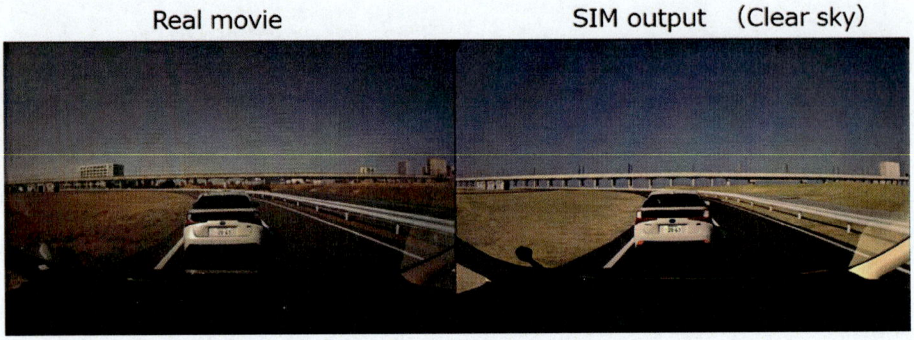

Fig. 5. Example of camera model simulation result

2.2.2 Millimeter-Wave Radar Model

Millimeter wave radars are the most difficult sensors for modeling. Three reflection models are defined and used according to the behavior of radio waves at reflective targets. The Physical Optics approximation is used as the scatter model for small objects such as cars and people, while the Geometric Optics approximation is used for large objects such as buildings and road surfaces as the reflector model.

In addition, the Radar Cross-Section (RCS) model is used to shorten the analysis time, which is defined for each object in advance, and assigns them to the objects in a combined scenario.

As the result, we get good coincidence with the real environment. (Fig. 6)

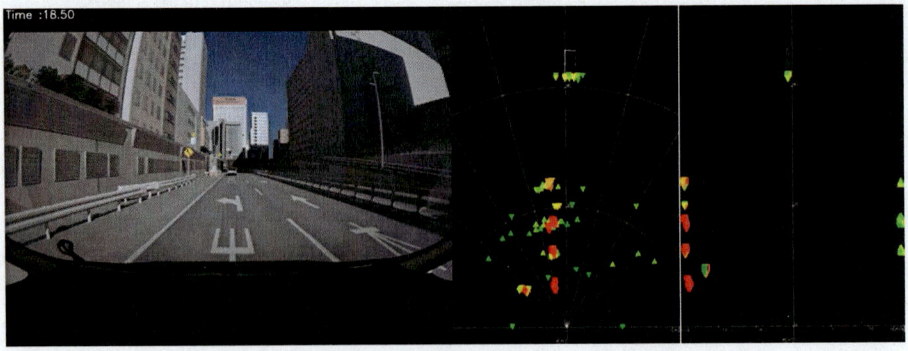

Fig. 6. Example of millimeter-wave radar model simulation result

2.2.3 LiDAR Model

LiDAR is relatively easier to model due to its directivity feature. It is possible to evaluate environmental disturbances such as background light through a 360° scan. In this project, we realized simulations that are highly consistent with the real world. (Fig. 7)

Fig. 7. Example of lidar model simulation output

2.2.4 Real Physics Based Approach

Our physical modeling framework follows a process circle, which is composed of System Identification, Simulation Modeling, Experimentation, Correlation, and Gap Analysis. (Fig. 8).

This process circle leads to models' consistency. Basic verifications were conducted at the laboratory level and at the providing ground level. This includes static and dynamic verification.

Based on these results, highly consistent simulations were realized for cameras and LiDAR, and progress was made in validating the consistency of millimeter wave radar simulations with particular objects.

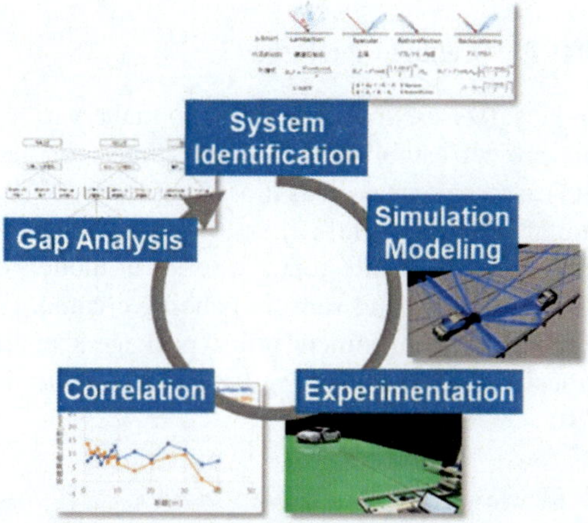

Fig. 8. Real physics based approach

2.3 Environmental Model

For the validation of ADS safety, it is necessary to define a virtual space as a unit, which can be called a package scenario according to the purpose, and to improve the level of reliable safety assessment in this scenario package unit.

Now we are discussing about sensing weakness scenarios with relevant members. The reproduction of these scenarios, which can be called "Virtual Proving Ground", can be extended to system validations that consider the effects of the environment such as rain, fog, and westering sun in a virtual space.

To apply reflection characteristics to object surfaces, this project focused on physical experiments and measurements of objects, which were then incorporated into the simulation models.

DIVP® is continuing to measure and validate the properties of high-priority asset models in order to help model sensing weakness conditions under real-world environmental conditions. (Fig. 9).

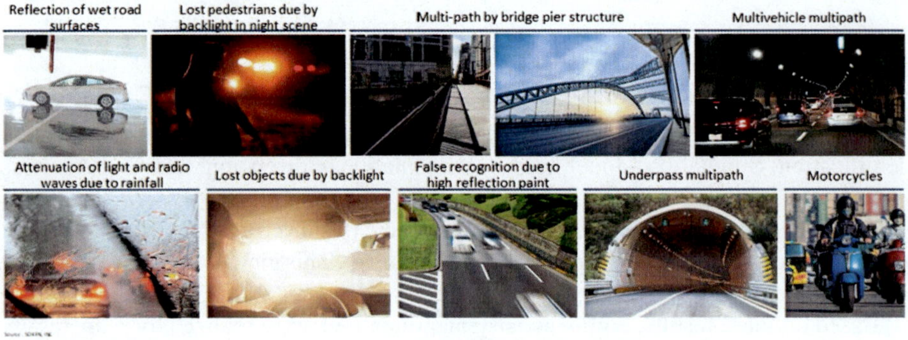

Fig. 9. Asset model to validate sensor weakness condition

2.4 Roadmap for DIVP® Scenario Package

In order to contribute the safety assurance, it needs to make various scenarios for evaluation. Therefore, this project established two milestones for the construction of an assessment scenario: (1) assessments such as those adopted by NCAP (New Car Assessment Program) or similar programs, and (2) assessments using community models of FOTs in Odaiba. Then, under these milestones, a series of models consisting of driving environments, spatial propagation, and sensors is being created. For this purpose, it is necessary to determine virtual environment units (package scenarios) in line with the targets and to continuously raise the level of reliable safety assurance for these scenario package units. (Fig. 10)

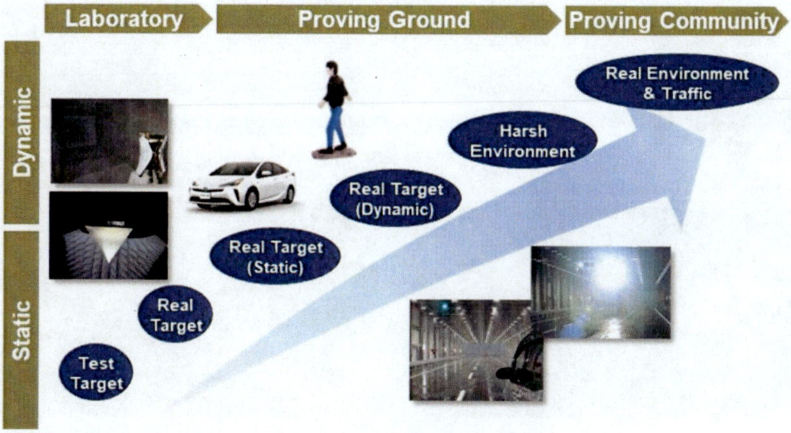

Fig. 10. Roadmap for DIVP® scenario packages

2.5 Application Examples

In order to contribute to safety validation for industrial players, DIVP® conducted a joint application trial with "AD-URBAN" project [2] also in SIP-adus. During the collaboration, DIVP®'s space design output injected into AD-URBAN's Fusion & Automated Driving control function to check the connectivity and validated the use case of Simulation based ADS safety validation. Through the collaboration, DIVP® can indicate the reproduction of some sensing weakness space designed models. Two examples are shown below.

2.5.1 Traffic Light Recognition Malfunctions' Model Due to Building Reflections that are Difficult to Reproduce in Actual Driving

First is the traffic light recognition malfunctions model due to building reflections. The conditions that make it difficult to detect traffic lights with this camera are the sunshine reflecting off buildings during the 10-min period from sunrise to early morning, and the relative positions of the vehicle's position, the building's reflection point, and the traffic lights. Since DIVP® simulation can create a physical model that realistically simulates the reflections under such conditions, it is very effective in developing camera recognition for ADSs. (Fig. 11)

2.5.2 Robustness Validation of the Localization Algorithm of AD-URBAN's ADS Using DIVP® Simulation

The second example is the robustness validation of the localization algorithm of AD-URBAN's ADS using DIVP® simulation. In the ADS of AD-URBAN, they are developing a highly accurate localization technology – even in urban areas. The basic part of the system is map matching technology using the LiDAR and other sensors as well as two-dimensional road pattern images created as ortho-maps from the ego-vehicle's

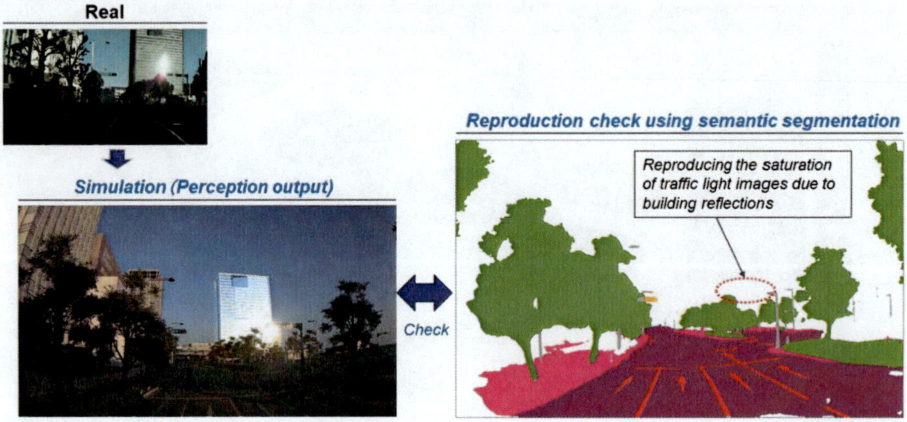

Fig. 11. DIVP® Simulation reality; reproduction of recognition failure conditions that are difficult to reproduce in actual driving

LiDAR and other sensors. In order to validate the robustness of the localization algorithm, the DIVP® simulation created a virtual space model having many parked cars in a row, which significantly reduces the amount of road surface features. Even under such severe conditions, the localization algorithm of AD-URBAN was found to be highly accurate. (Fig. 12).

DIVP® has an advantage of "difficulty scenario testing availability" for ADS safety as mentioned above.

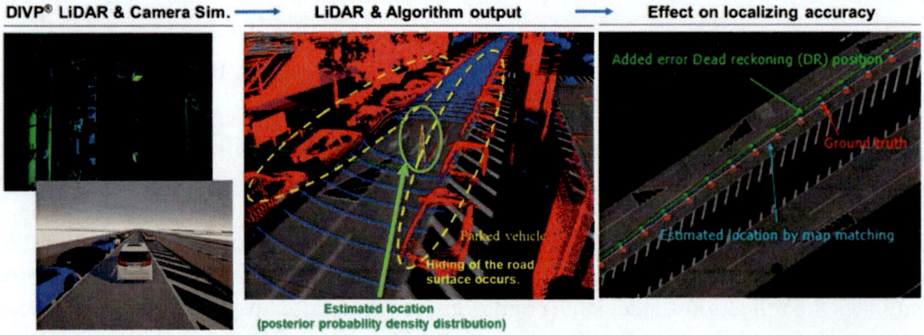

Fig. 12. Robustness validation of the localization algorithm on parked cars' scenario using DIVP® simulation

2.6 International Cooperation: Japan-Germany Collaborative VIVID Project

"How safe is safe enough?" and "How realistic is realistic enough?" are international questions. So DIVP® and VIVALDI of German national project launched a joint project named VIVID from November 2020. We share the information and conduct discussions about Test chain, Multi-sensor platforms, interfaces, and so on.

Through VIVID, the standardization of automated driving safety assurance systems and interfaces is currently in progress. (Fig. 13)

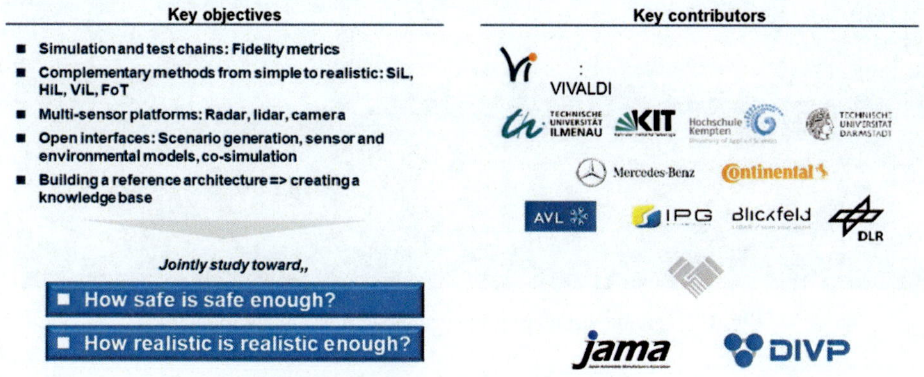

Fig. 13. VIVID implementation structure

3 Conclusions

This project has promoted industry-academic-government collaborative research and development as a part of the SIP-adus program. Utilizing the outcome of a series of models consisting of driving environments, spatial propagation, and sensors, this project would also like to enable connectivity with other simulators and provide a fundamental technological base to enable the efficient and widespread implementation of increasingly complex automated driving safety validations. DIVP® pursues to create a precise digital twin of real-world environments which can help raise consumer acceptance of the safety of automated driving while also helping to accelerate the social implementation of automated driving.

4 Next Steps

For the validation of ADS safety, it is necessary to define a virtual space as unit. We have created an environmental model of Odaiba in the Tokyo water front area. (Fig. 14) In this virtual space, real environmental factors will be combined to validate the sensor weakness scenario. Many stakeholders have an interest in the DIVP® simulation platform. Now they will monitor and experience our simulation platform. After this monitoring test, we plan to commercialize our DIVP® simulation platform in the market. We hope that our simulation platform will contribute to the safety assurance of ADS as well as a safe and smooth Automated driving society very soon.

Fig. 14. Odaiba virtual proving ground (Environment model)

References

1. Inoue, H.: Development of driving intelligence validation platform (DIVP®) for automated driving safety assurance. In: SIP 2nd Phase Automated Driving for Universal Services Mid-Term Results Report (2018–2020), 3rd chapter, pp. 89–94 (2021)
2. Suganuma, N., et. al.: Research on the recognition technology required for automated driving technology (Level 3 and 4). In: SIP 2nd Phase Automated Driving for Universal Services Mid-Term Results Report (2018–2020), 3rd chapter, pp. 79–85 (2021)

The Great Promise of AV - But it Needs to be Done Properly

Robert L. Sumwalt(✉)

National Transportation Safety Board, Washington, D.C, USA
aviatrends@gmail.com

Abstract. Autonomous vehicle (AV) technology has great promise. It has the potential to significantly reduce fatalities and injuries caused by roadway crashes, and it has the potential to eliminate roadway congestion. The devil is in the details, however, and to achieve these benefits, the development and implementation must be done properly. The author lays out thoughts that manufacturers and regulatory authorities may consider – thoughts that are intended to minimize unintended consequences and to ensure proper introduction of AV technology.

1 Introduction

Your work is changing the world. Let me repeat that: YOUR WORK IS CHANGING THE WORLD!. Literally.

I know you all understand the possible life-saving potential that autonomous road vehicles can have. There are around 1.3 million roadway-related fatalities each year around the globe, and nearly 40,000 fatalities and numerous injuries each year right here in the US. AV technology, if implemented properly, shows great promise for reducing many, if not most, of these tragic deaths and injuries.

There's also the potential for great efficiency gains. A few weeks ago – the day after I retired from a 15-year career with the National Transportation Safety Board (NTSB) – I packed up my car and headed home to my family home in South Carolina. I got on the major north-south artery, Interstate I-95, and for the first three miles, everything went smoothly. I was traveling at the posted speed limit. I glanced at my watch and figured I should be home a bit earlier than I had initially planned. Then it happened. A major highway backup – something that isn't unusual through much of urban and suburban areas – and especially not unusual for I-95 (as anyone who lives in northern

Robert Sumwalt was appointed to serve on the National Transportation Safety Board (NTSB) in 2006 by President George W. Bush. He was reappointed by President Barack Obama, and again by President Donald Trump. In 2017, President Trump appointed him as chairman of the NTSB. He stepped down from his position on the Board in June 2021.

The NTSB is an independent federal agency of the US government. It is charged by Congress to investigate transportation accidents, determine probable cause, and issue safety recommendations to prevent future accidents. The agency's statutory responsibility is to investigate all civil aviation accidents that occur in the US, as well as selected accidents in rail, marine, highway, and even pipeline accidents.

G. Meyer and S. Beiker (Eds.): ARTSymposium 2021, LNMOB, pp. 23–33, 2023.
https://doi.org/10.1007/978-3-031-11112-9_3

Virginia is keenly aware.) The cars slowly inched forward and eventually clearing that bottleneck – one that was apparently due to a minor "fender-bender" crash. This process of stop-and-go traffic continued off and on for the next 100 miles. Sometimes due to minor traffic crashes, and sometimes due to highway construction, sometimes due to merging traffic, and sometimes for no apparent reason at all. Through AV technology, many of these bottlenecks could be anticipated, and the necessary merging, or possibly rerouting could occur before they become time-wasting bottlenecks.

Until these increased safety and efficiency gains are achieved, the testing and deployment of AV systems requires appropriate safeguards and close interactions between Federal agencies, state and local governments, and industry.

In testimony to the US Senate in November 2019, I highlighted the great promise that AV technology offers, but I stated that I needs to be done properly. (Sumwalt 2019). However, the devil is in the details, so what does "properly" look like in my mind. In the following sections, I'll provide some thoughts for industry and regulatory bodies to consider.

2 Industry Considerations

My advice to industry is that, just because something *can* be automated or designed with fancy technology, doesn't necessarily mean it *should* be so designed. Granted, there are many brilliant people developing technological advances, and they can figure out how to do just about anything. Thanks to these bright thinkers, we probably have more computing capability on our cell phones than NASA had when they launched men to the moon. But, sometimes more technology isn't always the best solution.

Have you driven a rental car where you can't figure out how to turn on the radio or adjust the temperature? I experience this whenever I take my car to the dealership for maintenance and they loan me the latest and greatest (and usually the most expensive) model. To accomplish something that should be simple, such as turning on the car radio or adjusting the car's temperature, you need to figure out how to use console trackpad to select something on a panel-mounted multi-function display. Sometimes a traditional knob is more user-friendly and perhaps requires less attention to operate. Why, then, do auto manufacturers insist on higher tech? Perhaps it's just because they can.

An extreme case of technology outpacing human comprehension occurred on August 21, 2017, when US Navy destroyer, *USS John S. McCain*, veered abruptly into the path of a cargo vessel that was traveling on a parallel course. The bow of the tanker rammed into the left side of the *McCain*, penetrating a crew birthing area. Ten sailors were killed (NTSB 2019a). NTSB's investigation learned that the year prior to the accident, the *McCain* was upgraded with a new Ship Control Console (SCC). While most ships, including most other US Navy destroyers, have conventional mechanical throttle levers to control engine speed, this function on the *McCain* was controlled by a touch-screen flat-panel display (see Fig. 1). The left and right throttles control engine speed, which, in turn, changes the speed of the respective left and right propellors.

To increase or decrease engine speed, a sailor would simply use the touch-screen to move the electronically depicted throttles. In a typical operation, both throttles "ganged" together, so that moving one throttle would simultaneously move the other. However,

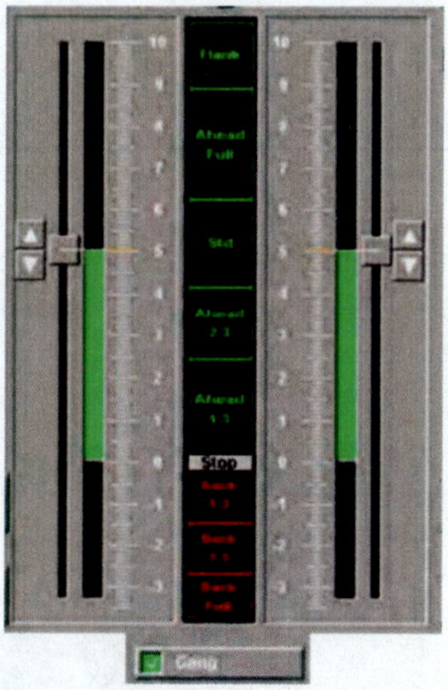

Fig. 1. Throttle touch-screen display onboard USS *John S. McCain.*

when the GANG checkbox at the bottom of the display is unchecked, each throttle can be moved individually.

Due to confusion on the *McCain's* bridge, unbeknownst to the bridge crew, the throttles became un-ganged. Therefore, when the sailor reduced engine speed on the left throttle, the right throttle remained at its previous position. The asymmetric engine speeds (higher propellor thrust on the right prop and less thrust for the left propellor) set up the abrupt turn to the left.

There were extenuating factors revealed in the NTSB's investigation, such as lack of adequate training and insufficient understanding of these systems. However, the NTSB noted in the final report of the accident that mechanical throttles, unlike the touchscreen throttles used on the *McCain*, "… provide complementary information to an operator: direction, force, and the ability to confirm either visually or by touch whether the throttles are ganged and working in unison. Mechanical throttles are used in aviation and on most vessels still operating in the Navy. They are often preferred over touch-screen displays as they provide both immediate and tactile feedback to the operator" (NTSB, 2019a, p. 33).

Since the accident, the Navy has begun replacing the newer SCC's, which includes touch-screen flat panel displays, with conventional mechanical throttles. So, back to the point: advanced technology isn't always the answer. Just because something can be automated or advanced technologically speaking, doesn't mean that it should be used.

2.1 Unintended Consequences

One thing to be on guard for is unintended consequences. James Reason,[1] once told me, "For every management decision, there is a potential downside that must be managed."

[1] Dr. Reason is perhaps best known for his development of the "Swiss cheese model," which illustrates how multiple layers of defense (the layers of Swiss cheese) can be breached (the

An example of well-intentioned people making decisions with consequence that were not fully understood is this crash of an experimental AV in 2018. On the evening of March 18, 2018, an automated test vehicle, based on a modified 2017 Volvo XC90 sport utility vehicle (SUV, see Fig. 2), struck a female pedestrian walking across the roadway in Tempe, Arizona. The SUV was operated by the Advanced Technologies Group (ATG) of Uber Technologies, Inc., which had modified the vehicle with a proprietary developmental automated driving system (ADS). A female operator occupied the driver's seat of the SUV, which was being controlled by the ADS. Although nighttime conditions existed, the road was partially illuminated by street lighting (NTSB 2019b).

Fig. 2. The accident vehicle involved a Volvo SUV modified with proprietary developmental automated driving system.

The SUV was completing the second loop on an established test route that included the section of road where the collision occurred. The vehicle had been operating about 19 min in autonomous mode—controlled by the ADS—when it approached the collision site in the right lane at a speed of 45 mph, as recorded by the ADS. About that time, the pedestrian began walking across the street where there was no crosswalk, pushing a bicycle by her side.

The ADS detected the pedestrian 5.6 s before impact. Although the ADS continued to track the pedestrian until the crash, it never accurately classified her as a pedestrian or predicted her path. By the time the ADS determined that a collision was imminent, the situation exceeded the response specifications of the ADS braking system. The system design precluded activation of emergency braking for collision mitigation, relying instead on the operator's intervention to avoid a collision or mitigate an impact.

Video from the SUV's inward-facing camera shows that the operator was glancing away from the road for an extended period while the vehicle was approaching the pedestrian. Specifically, she was looking toward the bottom of the SUV's center console, where she had placed her cell phone at the start of the trip. The operator redirected her gaze to the road ahead about 1 s before impact. ADS data show that the operator began steering

holes in the cheese). When all of the layers have holes, an accident or incident can occur because the defenses are absent.

left 0.02 s before striking the pedestrian, at a speed of 39 mph. The pedestrian died in the crash. The vehicle operator was not injured. Toxicological tests on the pedestrian's blood were positive for drugs that can impair perception and judgment.

NTSB determined that ATG had deactivated the Volvo's forward collision warning system and the Automated Emergency Braking (AEB). Their rationale for doing seemed reasonable: they didn't want those systems interfering with the ADS that they were testing. However, in doing so, there were unintended consequences that ATG must not have fully appreciated. NTSB's post-crash simulations showed that the FCW system would have alerted the driver 2.5 s before impact, and that the AEB would have activated 1.4 s before impact. Assuming no response from the driver, and considering that only AEB activation, the SUV was predicted to avoid a collision with the pedestrian in 17 out of 20 variations of the pedestrian's movement. In the other three variations, the AEB reduced the impact speed to less than 10 mph.

NTSB's report stated: "The Uber Advanced Technologies Group's deactivation of the Volvo forward collision warning and automatic emergency braking systems without replacing their full capabilities removed a layer of safety redundancy and increased the risks associated with testing automated driving systems on public roads" (NTSB 2019b, p. 57).

Another finding of the investigation was that "Uber Advanced Technologies Group did not adequately recognize the risk of automation complacency and develop effective countermeasures to control the risk of vehicle operator disengagement, which contributed to the crash" (NTSB 2019b, p. 58).

One way of possibly minimizing unintended consequences is to have those in the organization – those who were not involved in developing the systems and subsystems -- to challenge assumptions and preliminary decisions. We all know that our human brains have a left side and a right side. The right side is usually associated with the "creative" side of the brain, while the left side is the more "logical" side. Using the analogy of the human brain, I'm proposing that AV developers not only must have a creative side of the organization (the brain's right side, if you will), but they also have something akin to the brain's left side. This part of the organization acts as devil's advocate to poke holes in proposed decisions to serve as a check and balance on those creative decisions made by the "right side" of the organization.

3 Considerations for Regulatory Authorities

Regulatory authorities have a critical role. On one hand, they should ensure that proper standards are developed and adhered to; on the other hand, they shouldn't be overly prescriptive or take drastic measures that stifle innovation. That's a hard act to balance, because usually the technology designers move much faster than a regulatory body can.

While there is often a desire to jump directly to the end of the technological spectrum—highly automated "self-driving" vehicles—it is imperative that regulators and policy makers do not ignore the risks associated with partial driving automation systems currently being operated on our highways.

Until the full benefits of AVs may be realized, the testing of developmental ADS—with all its expected failures and limitations—requires appropriate safeguards when conducted on public roads. Unfortunately, there has been an absence of safety regulations and federal guidance regarding how to adequately evaluate an ADS, which has prompted some states to develop their own requirements for AV testing.

In the absence of federal ADS safety standards or specific ADS assessment protocols, we have ended up with a hodgepodge of legislative requirements, as many states in the US have begun legislating requirements for AV testing. The development of state-based requirements could be attributed to the concerns of many states about the safety risk of introducing ADS-equipped vehicles on public roads. The requirements vary. Some states, such as Arizona where the Uber ATG crash occurred, impose minimal restrictions. Other states have established requirements that include a more in-depth application and review process. In the Tempe crash investigation, we determined that Arizona's lack of a safety-focused application-approval process for ADS testing at the time of the crash, and its inaction in developing such a process following the crash, demonstrate the state's shortcomings in improving the safety of ADS testing and safeguarding the public.

Currently, only around 60% of the states in the US have regulations pertaining to ADS testing, and within these states, the requirements for testing vary considerably. Furthermore, the existence of a regulation is not a sure indication of a comprehensive and safety-driven ADS testing policy. In fact, Arizona was one of the states that had some form of regulation pertaining to ADS testing, but, as stated previously, the safety application approval process was lacking. States that have no, or only minimal, requirements related to AV testing can improve the safety of such testing by implementing a thorough application and review process before granting testing permits.

The American Association of Motor Vehicle Administrators (AAMVA) has developed numerous model programs for motor vehicle administration, law enforcement, and highway safety in general. In May 2018, AAMVA published Jurisdictional Guidelines for the Safe Testing and Deployment of Highly Automated Vehicles. Although the guidance contains elements of ADS testing, the AAMVA document lacked specific guidance for developers on how to accomplish the included recommendations. The guidance did include a very important element—the need for jurisdictions to identify a lead agency and establish an AV committee to develop strategies for addressing AV testing. However, the guidance does not include recommendations requiring ADS developers to submit a safety plan and for the state's AV committee to review and approve such a plan.

Because states would benefit from adopting regulations that require a thorough review of ADS developers' safety plans, including methods of risk management, NTSB recommended that AAMVA encourage states to (1) require developers to submit an application for testing ADS-equipped vehicles that, at a minimum, details a plan to manage the risk associated with crashes and operator inattentiveness and establish countermeasures to prevent crashes or mitigate crash severity within the ADS testing parameters, and (2) establish a task group of experts to evaluate the application before granting a testing permit. Similar recommendations were also issued to the state of Arizona.

In the NTSB's investigation of the Tempe, Arizona crash, the Safety Board found that Arizona lacked a safety-focused application approval process for ADS testing at the time of the crash (NTSB 2019b).

In 2021 NHTSA published an Advanced Notice of Proposed Rulemaking (ANPRM) to solicit thoughts from stakeholders on the proposed framework for ADS (NHTSA

2020). In response, on February 1, 2021, as then-chairman of the agency, I signed a letter on behalf of NTSB with the agency's response (Sumwalt 2019). We stated that DOT and NHTSA must first develop a strong safety foundation that will support the framework envisioned for AVs of the future. That foundation should include sensible safeguards, protocols, and minimum performance standards to ensure the safety or motorist and other vulnerable road users. We also called for the standardization of AV data collection to better understand automated control systems, along with a requirement for safety critical information to be available and evaluated for developmental ADSs, and the development of performance standards to evaluate driver engagement. Additionally, we suggested NHTSA improve oversight of systems that may operate outside a vehicle's operational design domain (ODD), and the incorporation of more robust collision avoidance test procedures into the New Car Assessment Program (NCAP).

3.1 Operational Design Domain Restrictions

SAE J3016 discusses the need for manufacturers to accurately describe AV features and clearly define the level of driving automation and its capabilities, but also its operational design domain—the conditions in which the driving automation system is intended to operate. Examples of such conditions include roadway type, geographic location, clear roadway markings, weather conditions, speed range, lighting conditions, and other manufacturer-defined system performance criteria or constraints. Tesla, for example outlined many operating conditions and limitations based upon the Autopilot partial automation system design, such as that it is (1) designed for use on highways with a center divider, (2) designed for areas with no cross traffic and clear lane markings, (3) not for use on city streets or where traffic conditions are constantly changing, (4) not for use on winding roads with sharp curves, and (5) not for use in inclement weather conditions with poor visibility.

Despite communicating to owners and drivers these operating conditions and limitations, Tesla Autopilot firmware does not restrict the system's use based on functional road classification. Essentially, the system can be used on any roads with adequate lane markings. This situation allows a driver to activate driving automation systems at locations and under circumstances for which their use is not appropriate or safe, such as roadways with cross traffic. The Tesla Model S in the Williston, Florida, crash collided with a tractor-trailer combination vehicle crossing an uncontrolled intersection on a nonlimited access highway (NTSB 2017).

Partial AV operation on nonlimited access highways presents challenges with the detection of crossing vehicles, pedestrian and bicycle traffic, and traffic controls at intersections, such as red traffic lights. As a result, NTSB concluded that, if AV control systems do not automatically restrict their own operation to those conditions for which they were designed and are appropriate, the risk of driver misuse remains. NTSB recommended that Tesla and other manufacturers of Level 2 automation: Incorporate system safeguards that limit the use of automated vehicle control systems to those conditions for which they were designed. (NTSB recommendation H-17-41) Five automobile manufacturers responded to this recommendation with steps they were taking to mitigate operation under conditions for which they were designed. Tesla, however, advised us that operational design limits are not applicable to Level 2 driver assist systems, such as Autopilot, because the driver determines the acceptable operating environment.

Tesla vehicles continue to be involved in crashes with Autopilot engaged in operating areas outside the intended roadway operational design domain. In March 2019, in Delray Beach, Florida, a fatal crash involving a 2019 T Model 3 occurred under circumstances very similar to the Williston, Florida, crash (NTSB 2020). The Delray Beach highway operating environment, like the cross-traffic conditions in Williston, was outside the Tesla Autopilot system's operational design domain.

Today's Level 2 partial driving automation systems can assess the vehicle's location and current roadway type or classification, and determine whether the roadway is appropriate to the system's operational design domain. Following the Williston crash, NTSB made a recommendation to NHTSA to address this vital safety concern. We recommended that NHTSA "Develop a method to verify that manufacturers of vehicles equipped with Level 2 vehicle automation systems incorporate system safeguards that limit the use of automated vehicle control systems to those conditions for which they were designed" (NTSB recommendation H-17-38). In response to Safety Recommendation H-17-38, NHTSA wrote the following: "The agency has no current plans to develop a specific method to verify manufacturers of vehicles equipped with Level 2 systems incorporate safeguards limiting the use of automated vehicle control systems to those conditions for which they were designed. Instead, if NHTSA identifies a safety-related defect trend in design or performance of a system, or identifies through its research or otherwise, any incidents in which a system did not perform as designed, it would exercise its authority as appropriate."

NTSB believes that NHTSA's reactive, rather than proactive, safety position is misguided, and the agency should take immediate action to verify that manufacturers are incorporating operational domain design safeguards into their systems, and therefore, classified their response as "Open—Unacceptable Response."

3.2 Monitoring an AV Driver's Level of Engagement

Based on system design, in an SAE-defined Level 2 partial automation system, it is the driver's responsibility to monitor the automation, maintain situational awareness of traffic conditions, understand the limitations of the automation, and be available to intervene and take over for the partial automation system at any time. In practice, however, drivers are poor at monitoring automation and do not perform well on tasks requiring passive vigilance.

Research shows that drivers often become disengaged from the driving task, both for momentary and prolonged periods during automated phases of driving. In the Williston, Florida, crash, NTSB found that the driver was disengaged from supervising the Autopilot partial automation. Tesla assesses the driver's level of engagement by monitoring driver interaction with the steering wheel through changes in steering wheel torque. In the Williston accident, when Autopilot was active prior to the crash, the system detected that the driver applied steering wheel torque only 2% of the time. Because Tesla uses steering wheel torque as a metric of driver engagement, the low percentage of driver applied torque in the Williston crash indicated a highly disengaged driver. This measure of driver engagement, however, is misleading. Because driving is a highly visual task, a driver's touch or torque of the steering wheel may not accurately indicate that he or she is fully engaged with the driving task. Simply checking whether the driver has placed a hand on the steering wheel gives little indication of where the driver is focusing his or her attention.

Following NTSB's Williston investigation, NTSB concluded that the way the Tesla Autopilot system monitored and responded to the driver's interaction with the steering wheel was not an effective method of ensuring driver engagement. As a result, NTSB recommended that six manufacturers of vehicles equipped with Level 2 driving automation systems "Develop applications to more effectively sense the driver's level of engagement and alert the driver when engagement is lacking while automated vehicle control systems are in use" (NTSB recommendation H-17-42).

In response to Safety Recommendation H-17-42, five of the six manufacturers responded with actions they were taking to monitor a driver's level of engagement. Tesla was the only manufacturer that did not officially respond. Because the operational design of partial driving automation systems requires an attentive driver as an integral system element, we will continue to advocate for manufacturers' improved monitoring of driver's level of engagement while supervising automation.

3.3 Event Data Recorders for Automated Vehicles

Title 49 CFR Part 563 sets forth requirements for data elements, data capture and format, data retrieval, and data crash survivability for event data recorders (EDRs) installed in light vehicles manufactured on or after September 1, 2012. The regulation did not mandate the installation of EDRs in light vehicles; rather, if the vehicle manufacturer chose to install an EDR, the regulation defines the format and specifies the requirements for providing commercially available tools and the methods for retrieving data from the EDR in the event of a crash. On December 13, 2012, NHTSA issued a notice of proposed rulemaking (NPRM) that proposed a new Federal Motor Vehicle Safety Standard (FMVSS) mandating that an EDR that meets 49 CFR Part 563 requirements be installed on most light vehicles. On February 8, 2019, NHTSA withdrew the NPRM because the agency determined that a mandate was not necessary. NHTSA's internal analysis showed that, for model year 2017, 99.6% of new light vehicles sold were equipped with EDRs that met Part 563 requirements. NHTSA added that, given the near universal installation of EDRs in light vehicles, it no longer believed that the safety benefits of mandating EDRs justified the expenditure of limited agency resources.

In withdrawing the final rule, NHTSA said that it would continue its efforts to modernize and improve EDR regulations, including fulfilling the agency's statutory mandate to promulgate regulations establishing an appropriate recording duration for EDR data to "provide accident investigators with vehicle-related information pertinent to crashes involving such motor vehicles." Because 49 CFR 563 data recording requirements codified more than a decade ago are very limited (only 15 data elements require reporting), NHTSA stated that it is actively investigating whether the agency should consider revising the data elements covered by Part 563 to account for advanced safety features.

In recent Tesla crash investigations, NTSB was able to retrieve data from the EDR, but the EDR data recorded did not address the partial driving automation system's activation or engagement. As a result, NTSB used other proprietary manufacturer data to interpret the automation system's functionality, but this type of data is not available on many vehicles operating with these systems today. Further, there are currently no commercially available tools for an independently retrieving and reviewing any non-EDR vehicle data, and other manufacturers of vehicles with driving automation systems control access to the postcrash proprietary information associated with their vehicles.

As more manufacturers deploy driving automation systems on their vehicles, to improve system safety, it will be necessary to develop detailed information about how the active safety systems performed during, and how drivers responded to, a crash sequence. Manufacturers, regulators, and crash investigators all need specific data in the event of a system malfunction or crash. Recorded data can be used to improve the automated systems and to understand situations that may not have been considered in the original designs. Crash reconstructionist need effective event data to conduct valid and productive investigations involving vehicles using AV control systems. Further, data are needed to distinguish between automated control actions and driver control actions.

Following the Williston crash, NTSB made a recommendation to the US Department of Transportation (DOT) regarding the need to define data parameters necessary to understand AV control systems and two recommendations to NHTSA to define a standard reporting format and to require manufacturers equipped with driving automation systems to report incidents, crashes, and vehicle miles operated with the systems enabled.

Specifically, NTSB called on DOT to "Define the data parameters needed to understand the automated vehicle control systems involved in a crash. The parameters must reflect the vehicle's control status and the frequency and duration of control actions to adequately characterize driver and vehicle performance before and during a crash" (NTSB recommendation H-17-37).

NTSB also called for NHTSA to "Use the data parameters defined by the U.S. Department of Transportation in response to Safety Recommendation H-17-37 as a benchmark for new vehicles equipped with automated vehicle control systems so that they capture data that reflect the vehicle's control status and the frequency and duration of control actions needed to adequately characterize driver and vehicle performance before and during a crash; the captured data should be readily available to, at a minimum, NTSB investigators and NHTSA regulators" (NTSB safety recommendation H-17-39).

NTSB also recommended that NHTSA "Define a standard format for reporting automated vehicle control data and require manufacturers of vehicles equipped with automated vehicle control systems to report incidents, crashes, and vehicle miles operated with such systems enabled" (NTSB recommendation H-17-40).

4 Closing

In wrapping it up, your work had the potential to save the world. But, as stated, it must be done properly. That means the industry needs to protect against unintended consequences. It means that the regulatory authorities need to do their jobs and make sure products are introduced in a manner that truly improves safety, but it must be done in a way that doesn't stifle innovation with overburdensome requirements. Those two regulatory responsibilities will be difficult to balance, but doing so is essential.

References

NHTSA: National Highway Traffic Safety Administration (NHTSA) advance notice of proposed rulemaking (ANPRM). In: Framework for Automated Driving System Safety, published at 85 Federal Register 78058 (2020). https://www.regulations.gov/comment/NHTSA-2020-0106-0617. Accessed 3 Dec 2016

NTSB: Highway accident report: collision between a car operating with automated vehicle control systems and a tractor-semitrailer truck Near Williston, Florida, NTSB/HAR-17/02. Washington, DC: Author (2017). . https://www.ntsb.gov/investigations/AccidentReports/Reports/HAR1702.pdf. Accessed 7 May 2016

NTSB: Marine Accident Report. Collision between US Navy Destroyer John S. McCain and Tanker Alnic M.C.. Singapore Strait, 5 Miles Northeast of Horsburgh Lighthouse. August 21, 201 (2019a). https://www.ntsb.gov/investigations/AccidentReports/Reports/MAR1901.pdf. Accessed 21 Aug 2019a1

NTSB. (2019b). Highway Accident Report: Collision Between Vehicle Controlled by Developmental Automated Driving System and Pedestrian. (NTSB Report No. NTSB/HAR-19–03). NTSB, Washington, DC (2019a). https://www.ntsb.gov/investigations/AccidentReports/Reports/HAR1903.pdf

NTSB: Highway Accident Report: Collision Between Car Operating with Partial Driving Automation and Truck-Tractor Semitrailer Delray Beach, Florida NTSB, Washington, DC. https://www.ntsb.gov/investigations/AccidentReports/Reports/HAB2001.pdf. Accessed 1 Mar 2019

Sumwalt, R.: Testimony Board Before the Committee on Commerce, Science, and Transportation United States Senate On Highly Automated Vehicles: Federal Perspectives on the Deployment of Safety Technology. Washington, DC. November 20, 2019. Retrieved from https://www.commerce.senate.gov/services/files/B8EF39B5-DE24-48AA-A870-B6CF8E0D5033

How Critical is Connectivity?

Katherine Kortum(✉)

Transportation Research Board, 500 5th Street NW, Washington, DC, USA
kkortum@nas.edu

Abstract. In preparing for widespread deployment of automated vehicles and shared mobility, it is important to understand the role of connectivity. This chapter discusses the importance of connectivity from the perspective of OEMs, other private sector representatives, and federal, state, and local government agencies. Key findings are that all stakeholders see great value in increased connectivity and that it is politically feasible, unlike other means of increasing transportation capacity, such as building more roads or congestion pricing. However, it is very easy to get caught up in "shiny object syndrome" with connectivity and forget about the purpose of implementing it: the human who is the end user. A number of research questions are outlined at the end of this chapter.

Keywords: Automated vehicles · Connectivity

1 Introduction

The 2021 Automated Road Transportation Symposium, organized by the Transportation Research Board (TRB), included a Breakout Session titled *Preparing for AVs and Shared Mobility: How Critical is Connectivity?* This session comprised a panel of experts discussing the importance of connectivity to automation. The session also served as an extension of workshops held by the TRB Forum on Preparing for Automated Vehicles and Shared Mobility over three months in spring 2021.

Automated vehicles are a frequent source of conversation in the transportation and technology worlds, but the connectivity required to link vehicles and infrastructure is far less discussed. This breakout session discussed how critical connectivity is to the many goals that society hopes to achieve through automation: equity, safety, mobility, accessibility, environmental and energy sustainability, and economic development. It provided thoughts from a variety of perspectives including original equipment manufacturers (OEMs), other private sector representatives, and federal, state, and local government agencies. Each gave their thoughts and invited audience responses on the importance and role of connectivity as it relates to automated vehicles.

The breakout session's objectives were to provide an understanding of the *current uses of connectivity* in the field and perspectives on how critical connectivity is to the success of automated vehicles from OEMs, automotive and telecommunications industries, users, and government agencies. Organizers wanted to establish an understanding of *the role of connectivity in ensuring safety* in mixed-flow environments that include

G. Meyer and S. Beiker (Eds.): ARTSymposium 2021, LNMOB, pp. 34–41, 2023.
https://doi.org/10.1007/978-3-031-11112-9_4

automated vehicles and other transportation services, including public transit and micro-mobility devices. They also sought to provide discussion about *how to use connectivity* in the short-term to achieve automation goals in the long term and *best practices* from state DOTs, the primary infrastructure owner/operators, on their deployment and management of connectivity options. And finally, as with all breakout sessions, organizers wanted to identify potential areas for *further research* and investigation.

2 Summary of Presentations

The session contained three separate panels and quite a bit of discussion following each one. The first panel focused on the perspective of OEMs with two representatives from industry organizations. A second panel provided user and industry perspectives from private companies and a research organization focused on safety. The third and final panel consisted of representatives of departments of transportation at the federal, state, and local levels.

2.1 OEM Perspectives

Ed Straub of SAE International opened the session, and he was followed by Carla Bailo of the Center for Automotive Research. Both acknowledged that all OEMs, generally known as vehicle manufacturers, are aware that many people would like to have access to their data. All OEMs also want to be able to keep their and their customers' data safe. The details of that balancing act can lead to tensions in the industry.

To achieve the full benefits of connectivity, infrastructure needs to be in place. Audi began to build connectivity into their vehicles (Hawkins 2022) but ended that practice. They could not turn on the connectivity very often because there was not enough infrastructure and equipment in the real world to make the system valuable to the driver. More and more connectivity is being developed and OEMs and local agencies both are gathering additional cloud data. Questions remain about how the data can be shared to the benefits of state DOTs, federal DOTs, and other partners to help with improving vehicle throughput and understanding driver behavior.

Both speakers agreed that no one, including developers, would refuse the concept of communication and connectivity between vehicles and infrastructure. However, this communication and connectivity is outside the control of manufacturers. Systems and vehicles are assessed on their current real-world performance. Developers therefore build vehicle systems to operate independently, without the infrastructure, because those systems are within their control and doing so allows them to determine the level at which their vehicle operates. In the absence of reliable data, developers turn any available data into just another sensor input, adding to the vehicle's overall perspective of its surroundings and to the amount of information available. Improving data standards, which will increase the reliability and availability of the data to allow it to come into a vehicle consistently, will be helpful to the industry as a whole (Bertoncello et al. 2021).

Others attending the session highlighted the costs of installation, writing code, and other connectivity system issues. OEMs have concerns about control of the connectivity systems and sensors and the reliability of the public sector in following through on infrastructure commitments. Changes in political power and administrations can cause significant shifts in what is built. The years-long lack of a federal infrastructure funding bill in the United States is also a related issue.

2.2 Industry and User Perspectives

David Yang of the AAA Foundation for Traffic Safety explained that discussions about the transportation system include three components: (1) infrastructure to support the transportation, (2) the vehicle carrying the people or goods, and (3) the user. When too much emphasis is on any one of these three pieces, the point of a complete transportation can be lost. Connectivity plays a critical role in bringing these three pieces together. Cooperative transportation systems, enabled by connectivity among the elements, enable users to be better aware of driving conditions and have more time to react and respond. Connectivity allows for clearer communication of intentions; currently, turn signals are one of the best signals of intention. Human factors experts often discuss mental models, which are a person's expectations for how a situation will go. Today, most people feel empowered inside their own vehicle, as though it is a closed-loop system. In order to truly improve safety and operations, people need to consider the transportation system as a shared system, in which, for example, the traffic signal is for all users and not only that driver. A more connected and cooperative system can allow the transportation industry to work towards its long-held goal of safe mobility.

Randy Iwasaki, now of Amazon but previously an executive at transit and surface transportation systems, described his move to a technology company as "trying to stake to where the puck is." Amazon, and particularly Amazon Web Services, focuses on data capture and connectivity. Integrating some of the company's tools into transportation helps to answer questions about providing access to customers where they are. New vehicles can generate 20 terabytes of data per hour (Magaia et al. 2021), requiring tools that can send information back to the host (the vehicle) within milliseconds. Those tools are available through technology companies like Amazon, helping to prepare corridors for the future so that they are safe to use. The industry has had a goal of reducing the traffic fatality rate for decades, but it has barely budged. Now the industry is starting to capture and analyze more data, especially at intersections, that can make it possible to save more lives. Amazon is agnostic to the type and technology of the connection, as long as they receive the data that the system needs. The company's artificial intelligence, machine learning, cameras, and other powerful analytics tools can work with systems and data anywhere, and bringing those tools back to transportation is the goal.

Barry Einsig of Econolite pointed out that the Defense Advanced Research Projects Agency (DARPA) off-road challenges, which were among the first major steps toward automated vehicles, were sixteen years ago, but the industry is still fundamentally using the same technology. The radar and LIDAR and cameras have all improved iteratively, but there have been no significant leaps forward. Improving these systems without adding connectivity is a "fool's errand." Considering both the cost and energy consumption of what is installed in vehicles today, the cost curves are going in the wrong direction and power consumption per vehicle is rising rapidly (Wadud et al. 2016). The cost to build a fully automated vehicle, at least one that is crashworthy, is at least a million dollars, and it will not happen without connectivity.

Mr. Einsig also described conversations about dedicated short range communications networks (DSRC) versus LTE/LTV versus 5G as missing the entire conversation. The network attached to these specific technologies is the concern, especially whether that network is public or private. It is an open question as to whether any private service will take on the liability needed for the data exchanges. Technology choices will be solved through business plans, but the transportation industry needs to focus today on use cases that are immediately deployable to reduce emissions, congestion, and costs of the system. We must develop a viable system that cities and states can use immediately, as they are the entities who are currently investing. The backbone of the network needs to precede the technology, because if the data cannot be exchanged over the network, a radio of any type is merely a beacon.

Finally, Jim Misener of Qualcomm pointed out the importance of connectivity and information across modes. He stressed that metropolitan planning organizations (MPOs) can use their preferred terminology for these needs, but that they have quite a few current use cases. These include vulnerable user road alerts, "do not pass" warnings, intersection movement assistance at blind intersections, blind curve and local hazard warnings, and work zones. It is also important to differentiate which data is useful to the vehicle and which to the user, and to deliver only the most critical information to the user. As Mr. Yang stated at the beginning of the session, this requires a focus on which component of the transportation system needs the alert or data stream and not overloading the system by providing alerts or data to unnecessary components.

2.3 Federal and Regional Perspectives

Federal and regional perspectives were provided by Kevin Dopart of the United States Department of Transportation (DOT), Kristin White of Minnesota DOT, Blaine Leonard of Utah DOT, Cathy McGhee of Virginia DOT, and Faisal Saleem of Maricopa County (Arizona).

Beginning at the federal level, Mr. Dopart specifically referred to low-latency vehicle-to-everything (V2X) efforts in his definition of connectivity. This is not on the critical path to widespread V2X, but achieving full safety and network efficiency benefits from automated vehicles will require low-latency V2X-style capabilities. Given the lack of control that the industry has over this sector, it is understandable that the industry will not invest very much yet until there is more stability. The United States DOT is funding performance testing on LTE V2X, including on adjacent band interference, with results to come in fall 2021 (USDOT 2021). The United States DOT will also conduct 5G device testing. The department will continue to work with state and local partners to develop solutions in the next year as they move away from DSRC implementation efforts.

Moving onto the state level, Ms. White of Minnesota DOT began by saying that connectivity is a central part of what connects communities. It is important to look at a systems approach, not only addressing specific goals, and state DOTs are the entity which needs to take a long-range approach. The American Association of State Highway and Transportation Officials (AASHTO) has a draft connectivity vision (AASHTO 2021). Their vision of the future is connected and automated, with redundant systems and looking at performance-based outcomes instead of specific technologies. Remaining technology-neutral while planning for the future is twofold. It understands that connectivity is ideal, but also does not assume that any vehicle will be connected. Minnesota is undertaking right-of-way partnerships and testing connected vehicle corridors, acknowledging that these might be sunk investments in the near future. The state believes that the backbone of a connected transportation system is in fiber, so it is trying to understand where the existing gaps are and which corridors are well-suited to public-private partnerships for further investment.

Overall, Minnesota believes that connectivity is an essential part of equity. States and other jurisdictions should prescribe the goals, not the means of achieving them, and leave it to the experts to figure out how to reach those goals. Many discussions center on big data, but sometimes small data is better; federal data exchanges can help to advance the work of right-sizing data needs. It is important to prioritize building public trust and consider the human factors impact of the work. If the public does not understand the systems and their importance, the transportation industry will go nowhere. Finally, Minnesota believes that the United States needs an equivalent of the European Union's General Data Protection Regulation, which went into effect in 2018.

Utah DOT's Blaine Leonard began by reminding the audience that the primary goal for both connectivity and automation is safety, although they may serve other purposes as well. He then laid out five use cases for why connectivity, along with automation is necessary to reach zero fatalities on the roads.

First is platooning, or cooperative strings of vehicles. Connectivity is necessary for these vehicles to talk to each other. Sensors can measure distances and changes in distances between vehicles, but they cannot measure or determine intent.

Second is intersection safety. Sensors on an automated vehicle can see a traffic signal in its current phase, but they cannot know what is coming. Giving vehicles information about signal phasing and timing is key to making many intersection applications function. Utah is working to make signal broadcasts reliable and consistent, but has not reached that point yet. A LIDAR installation at an intersection can provide additional information about non-sensored users in the area and inform a connected vehicle.

Third is road safety. Sensors on vehicles may be able to see ice that is on the road, but they will not be able to determine wind speed or quantities of blowing dust. Sensors on the road or nearby can make these determinations and inform the vehicles.

Fourth are work zones. It is important to provide information to a vehicle that may differ from what the high-definition map tells the vehicle. For example, Waymo vehicles have found themselves stuck in work zones, even very recently (Templeton 2021). This is not a simple problem, requiring the system to obtain real-time information from the field, compile it into an information package, and then submit it to the vehicle. OEMs and DOTs are all in agreement that work zones are a challenge.

Fifth and finally, connectivity provides invaluable redundancy to sensors to corroborate their information and/or provide additional information. If there is rain, glare, dust, or snow, the on-vehicle sensor may not be able to see well, and connectivity can add multiple observations from various vantage points.

Ms. McGhee of Virginia DOT pointed out that while OEMs are experts at building vehicles, there is a larger picture for mobility for all people. Virginia in particular has the country's third largest transportation system, and some of the most challenging geometry and highest crash rates are on state roads. Auxiliary components of the system, such as rest stations and bridges, were designed for human drivers and not automated systems. It is much easier to convert these to an automated environment with connectivity than without it. State partnerships with universities have allowed for a great deal of connected and automated vehicle testing, and the state has attempted to focus this testing on things it can influence as infrastructure owner/operators and that will help meet the statewide goals. Workers in work zones need to become part of this connected environment, because there are quite a few drivers who intrude upon work zones and the state wants to protect the workers who face danger from these drivers.

Mr. Saleem, of Maricopa County, began by describing the industry's initial attempts at connectivity in an era of 1G communications, and the process has remained much the same since that beginning. Maricopa County wants to consistently, reliably, seamlessly, and securely deliver quality digital products to consumers. It is hard to accomplish this regionally when so many jurisdictions are involved, but travelers move seamlessly across jurisdictions, and the County wants their experience to be smooth.

The County has found that three processes are essential to making connectivity work. The first is production flow and development of a product, whether that product is a message, a map, or a data frame. This requires a great deal of back-end processes. The County, for example, does not operate or manage work zones; construction groups do this work, and they need processes. These back-end processes are not always ready to be shared yet, or even in digital form. The second is the technology flow. In order to operate in the use cases that this session already discussed, the organization needs to have a handle on physical infrastructure, data, networks, presentation, applications, and more. The third essential process is business flow. Here, the operators need money, agreements, partnerships, lease/own agreements, organizations, and a workforce. Without all of these items, connectivity cannot function well.

3 Conclusions

3.1 Summary of Discussions

All panels and panelists agreed that connectivity is critical to full effectiveness of automation. There are many use cases where connectivity can solve problems that automation alone cannot.

The technology debates about DSRC versus 5G and decisions by the Federal Communications Commission decisions should not be the focus of connectivity discussions. Both items have received a great deal of attention, but they miss many other considerations: liability issues, data exchanges, and more.

Technology perspectives and DOT perspectives are on parallel tracks but will hopefully converge in the future.

No matter which specific technological solutions are deployed in coming years and decades, the processes we are developing and the learning curves we are currently scaling will be helpful.

3.2 Key Findings and Lessons Learned

Participants debated the competing visions and methods of DOTs and tech companies/OEMs. All are aimed at making the system better, but all take different routes to the eventual goal.

Technology companies can and do improve vehicles and safety; DOTs and other public agencies are broadly responsible for safety and quality of life of communities.

Infrastructure owner/operators, such as state DOTs, see great value in increased connectivity.

Connectivity is politically feasible, rarely resulting in significant policy opposition. This is in contrast to other means of increasing transportation capacity, such as building more roads, congestion pricing, and more, which face much more established opposition.

It is very easy to get caught up in "shiny object syndrome" with connectivity and forget about the purpose of implementing it: the human who is the end user.

4 Next Steps

The session resulted in a variety of research questions which can be addressed industry-wide as next steps.

How do predicted benefits of deployment of autonomous and shared vehicles change if connectivity is limited, e.g., what are the incremental benefits of deploying connected automated vehicles beyond deploying just (unconnected) automated vehicles?

Why is connectivity important from the end-user perspective, as opposed to the infrastructure and infrastructure perspective, and how do we convey that?

What use cases are required to be analyzed to create a full understanding of the potential benefits and other implications of connectivity?

What are the legislative and regulatory implications of connectivity? Similarly, what are the implications of public-private sector relationships, funding, and risk and liability assignments?

Finally, what needs to be done to enhance consistency regarding connectivity among the state DOTs, and between the DOTs and vehicle manufacturers? What standards are needed?

References

American Association of State Highway and Transportation Officials (AASHTO): AASHTO Policy Recommendations on Key Surface Transportation Priorities (2021). https://policy.transportation.org/wp-content/uploads/sites/59/2021/11/AASHTO-Policy-Recommendations-on-Key-Surface-Transportation-Priorities-V2-2021-10-08-FINAL.pdf

Bertoncello, M., Martens, C., Möller, T., Schneiderbauer. T.: Unlocking the full life-cycle value from connected-car data. McKinsey Center for future mobility (2021). https://www.mckinsey.com/industries/automotive-and-assembly/our-insights/unlocking-the-full-life-cycle-value-from-connected-car-data

Hawkins, A.J.: Audi is bringing 5G connectivity to its vehicles in 2024. The Verge, February 22 (2022). https://www.theverge.com/2022/2/22/22945823/audi-verizon-5g-connectivity-2024-v2x

Magaia, N., Mastorakis, G., Mavromoustakis, C., Pallis, E., Markakis, E.K.: Intelligent Technologies for Internet of Vehicles. Springer, Cham (2021). https://doi.org/10.1007/978-3-030-76493-7. ISBN: 3030764931

Templeton, B.: Waymo performs embarrassingly in construction cone situation. Forbes, May 14 (2021). https://www.forbes.com/sites/bradtempleton/2021/05/14/waymo-peforms-embarrassingly-in-construction-cone-situation/?sh=6b0d58e97009

Transportation Research Board (TRB): TRB Forum on preparing for automated vehicles and shared mobility: mini-workshop on the importance and role of connectivity. TRB E-Circular E-C247, February 14 (2019). https://onlinepubs.trb.org/onlinepubs/circulars/ec247.pdf

U.S. Department of Transportation (DOT): Test plan and test procedures summary: testing long-term evolution vehicle-to-everything (LTE-V2X) radio performance capabilities within the 5.9 GHz Safety Band (2021). https://www.transportation.gov/sites/dot.gov/files/2022-04/DOT%20LTE%20V2X%20Summary%20Overview%2007cDEC2021%20vFINAL.pdf

Wadud, Z., MacKenzie, D., Leiby, P.: Help or Hindrance? The travel, energy, and carbon impacts of highly automated vehicles. Transp. Res. Part A Policy Pract. **86**, 1–18 (2016)

Check for updates

Innovation Strategies and Funding Policies for Automated and Electric Road Mobility

Gereon Meyer(✉), Carolin Zachäus, and Jakob Michelmann

Department European and International Business Development, VDI/VDE Innovation + Technik GmbH, Steinplatz 1, 10623 Berlin, Germany
{gereon.meyer,carolin.zachaeus,jakob.michelmann}@vdivde-it.de

Abstract. While the automation and the electrification of cars at first appear as separate technology fields, their interdependencies provide synergetic and complementing effects at both the layers of technology and the applications. In addition to describing those effects, this chapter analyses how the integrated view of both paths is covered by industrial and academic research and innovation strategies, public funding programmes and collaborative projects in the European Union and its member states, taking Germany and Austria as examples. Furthermore, international benchmarks from outside Europe are presented, notably from the U.S, China, Japan and South Korea, and some future prospects are given. This chapter summarizes and concludes the activities of Task 29 „Electrified, Connected and Automated Vehicles" of the Technology Collaboration Programme Hybrid and Electric Vehicles (HEV-TCP) of the International Energy Agency (IEA), and reports on some outcomes of EU-funded Coordination and Support Actions in the domain of smart and sustainable road mobility.

Keywords: Electric vehicle · Connected and automated mobility · Autonomous driving · Electronic components and systems · Innovation Policy · Research funding

1 Introduction

Automation and electrification are the leading paradigms of the transformation of the automobile. Both of these trends describe innovation paths that have made significant progress during the last ten, twelve years. To date, their maturity as a product is different though. Electric vehicles have become commonplace and already show a market share of almost 20%, at least in Europe and China [1]. In contrary, for highly automated vehicles providing transportation services with reduced or no driver interaction, i.e. automation levels 3 or 4 according to SAE [2], the deployment has been limited to a number of demonstration and pilot projects in restricted operational design domains, so far. Nonetheless, further advancements can be expected for both innovation paths in the near future, be that in terms of technical functionality, supporting infrastructure of energy and data, or the transport system integration. Due to the potential coincidence of leap innovations, the interactions of automation and electrification have gotten into the

G. Meyer and S. Beiker (Eds.): ARTSymposium 2021, LNMOB, pp. 42–49, 2023.
https://doi.org/10.1007/978-3-031-11112-9_5

focus of engineers, planners, user and businesses, recently. Given the added benefits for road safety and climate neutrality, resp., relevant synergetic or complementary effects from the smart combination of the two paths have been supposed [3]. These are subject of current research innovation strategies and roadmaps and of public funding programmes in Europe, its Member States and beyond, though to a different extend.

2 Synergies and Complementarities

Already some basic considerations tell that the coincidence of technical progress in the automation with that in the electrification of road vehicles may lead to synergies and complementarities at both, the layer of the technical system and the layer of application:

In technical terms, links between the development and innovation processes in automation and electrification are likely since the electric and electronic architectures that control these two domains in the automobile are similar, if not the same smart systems: Whether the lidar sensors of a car perceive the road environment to avoid crashes with obstacles or other road users, or the cameras and radar sensors watch out for icy road surfaces or upcoming slopes to adjust the power request from the traction battery, are essentially just sensing tasks. Additionally, the power electronic converters driving the electric traction motor, or actuating the steering systems or the brakes, are almost equal actuation tasks. Moreover, electric drive systems are easy to control electronically. Their abundant electrical power would even support steering and braking by wire. After safety, assistance and propulsion functions have been provided by completely separate microcontrollers in the past, also the sensors and actuators and the respective flows of energy and data have been kept separately so far. However, with the trends towards more comprehensive electric and electronic architectures and vertical integration, these functions will in the future be controlled from a centralized car server, and accessed commonly either within their respective topographical zone in the car or vehicle-wide within the same domain. As a consequence, synergies may be due to a more intelligent and accurate control of the power flows and charging processes, cost and weight reductions caused by less complex wiring harnesses, and last but not least a greater fail operability and more flexible upgradability [4].

At application level, commonalities in the systematic nature of the operating environment for automation and electrification enlarge synergetic and complementary effects. Most prominently, an automated and electric vehicle may be operated more efficiently, and thus have a longer electric range, particularly if connected to other vehicles or receiving control signals from the infrastructure, e.g. traffic lights, that could help to avoid unnecessary acceleration and braking maneuvers. The combination and convergence of those two innovation paths may also define novel products, designs and services raising the usability of electric vehicles, e.g. quiet parcel delivery, waste collection or street cleaning at nighttime in cities. At the same time, electrification may give rise to new features of highly automated vehicles, e.g. always-on capabilities allowing software-updates over the air or power-intense robotic features in the building, maintenance or emergency domain. Once both technology fields are fully mature, one could imagine mobility-on-demand concepts of the sharing economy to benefit from automation and electrification by e.g. optimized fleet management, self-controlled wired or wireless charging and parking processes and better integration into the multimodal urban mobility system. On the longer term, additional synergies could arise from replacing the automobile's passive safety systems by active, automated ones, enabling a super light and extremely efficient electric vehicle that would be able to cover a very long range with a small, and again lightweight battery or fuel cell. This would lower the total-cost-of-ownership and further raise the usability of electric cars significantly, while reducing the consumption of energy and thus the greenhouse gas and noxious emissions of road transport.

On the other hand, rebound effects are a matter of concern: High-degree automation of vehicles may lead to a more intense use of them and thus increase their total energy consumption even though the vehicles may be more energy-efficient due to electrification. An extensive joint study by a number of U.S. National Laboratories recently concluded that connected and automated vehicles could potentially lead to a threefold increase or decrease of energy consumption in cars [5].

3 Research and Innovation Strategies in Europe

In view of the supposed coincidence of progress and innovation steps in automated mobility and electric vehicles, strategic research and innovation policies aim to anticipate the opportunities of a better-aligned innovation process and to identify and avoid imminent market failures. Therefore, industrial roadmaps and strategic research agendas, public research funding programmes and funded projects have covered the synergies and complementarities of the electrification and automation of road vehicles in recent years.

For the private and academic sectors in Europe, this is reflected in the strategic research agendas of the European Technology Platform on Smart Systems Integration (EPoSS) and the European Road Transport Research Advisory Council (ERTRAC) as well as in the innovation roadmaps of automotive research platforms and ecosystems at national levels, such as the Austrian Association for Advanced Propulsion System (A3PS) or the eNOVA Strategy Board Automobile Future in Germany. Even the Forschungsvereinigung Automobiltechnik (FAT), the research branch of the German Association of the Automotive Industry (VDA), has recommended more publicly funded research on exploiting the synergies of automation, connectivity and electrification in its latest roadmap [6].

For the public authorities in the European Union, the fundamental innovation paths in automation and electrification of road vehicles are outlined in the respective Strategic Transport Research and Innovation Agenda (STRIA) edited by the European Commission [7, 8], both making reference to the potential interdependencies with each other. In its efforts to put the EU on the path to transforming the mobility system of the future and bringing about the fundamental changes needed to achieve the objectives of the European Green Deal, the European Commission released its Smart and Sustainable Mobility Strategy in 2020 [8]. It includes the intention to launch two co-programmed partnerships under the ninth research framework programme, Horizon Europe, namely Towards Zero Emission Road Transport (2Zero) and Connected, Cooperative and Automated Mobility (CCAM).

The Strategic Research and Innovation Agendas of these two partnerships both again refer to each other, particularly regarding the synergies in terms of new vehicle and shared mobility concepts, transport system integration and user behavior studies [9, 10]. The synergies and complementarities at the technical systems level are considered by linking both, CCAM and 2Zero, to a third partnership of Horizon Europe, Key Digital Technologies (KDT), dedicated to Electronic Components and Systems (ECS) as common enabling technologies. Following the recommendation of the advisory group Lighthouse Mobility.E of the previous, eighth research framework programme, Horizon 2020, KDT describes the interaction with 2Zero and CCAM as a bidirectional strategic alignment process along the value chain. Therein, ECS are the building blocks enabling new functions for green and automated vehicles, while the concepts developed by CCAM and 2Zero define requirements for ECS. Potential synergies identified at the level of ECS include e.g. electric and electronic architectures for fail-safe power distribution and control within the vehicle, the functional safety and reliability of systems and cybersecurity, and intelligent control for power systems [11].

At European Member States level, e.g. the German Federal Government's action plan for automated mobility is pointing out the opportunities of automation for more efficient, sustainable and clean mobility solutions [12], while the Austrian Federal Government's research, technology and innovation (FTI) strategy in mobility contains separate pillars for climate-neutral propulsion systems and automation, digitization and connectivity. These pillars merge in their objectives to avoid, divert and improve mobility, in the mobility transition at urban and regional levels as well as in the key enabling technologies [13].

4 Funding Programmes and Projects in Europe

The European Commission is implementing its strategy for automated and electric road mobility through the research and innovation funding calls of the biannual work programmes of Cluster 5 "Climate, Energy and Mobility" of Horizon Europe. For the CCAM and 2Zero partnerships, this requires the European Commission to seek alignment with their counterparts from the private and the academic sector. The CCAM calls are part of the Destination 6 "Safe, resilient transport and smart mobility services for passengers and goods ", and the 2Zero calls belong to the Destination 5 "Clean and competitive solutions for all transport modes". While the 2021–22 work programme of Horizon Europe did not particularly address the synergies of automation and electrification, a joint call of CCAM, 2Zero and the European Mission "Climate-Neutral and Smart Cities", covering these synergies is foreseen for 2023. The outcomes expected from projects funded under this call include transferrable solutions for mobility of people and goods exploiting the combined potential of electrification, automation and connectivity. Except of pilots with automated shuttles, which normally are electric vehicles, e.g. in the SHOW project funded under Horizon 2020, there have not been any European research and innovation projects primarily looking into the synergies of automation and electrification, so far.

The situation is different at the level of EU member states, though: In Germany, funding calls dedicated to enabling technologies for automated and electric vehicles were published in the context of the Framework Programme Microelectronics 2016–2020 by the German Federal Ministry of Education and Research (BMBF) [14]. This includes the call on "Electronics for Autonomous, Electric Driving" (ELEKTRONOM) and the call on "Disruptive Vehicle Concepts for Autonomous Electric Mobility" (AutoDis), both from 2017. An example of the collaborative projects funded under these calls is UNICARAgil, which is clearly showing the potential synergies of electrification and automation at both levels, the electric and electronic architecture and the application side. It combines modular structures for agile, automated vehicles with disruptive concepts in hardware and software architecture and a modular platform with dynamic electric modules. By the end of the project, four prototype vehicles with different characteristics will be presented, ranging from a delivery van to an on-demand shuttle. In UNICARAgil, the synergetic features are a centralized E/E architecture, service-oriented software architectures allowing safe automation and efficient operation, and dynamic modules with 90 degree rotatable wheels, wheel motors and power electronics as actuators for new automated services. Further projects on technology development for automated and electric vehicles may be expected from the recently published funding calls on "Electronics and Software Development Methods for the Digitalization of Automobility" (MANNHEIM) and on "New Vehicle and System Technologies" that is part of the COVID-19 recovery programme, paragraph 35 on future investments for vehicle manufacturers and suppliers.

In Austria, dedicated funding calls have been published in the context of a programme Mobility of the Future by the Federal Ministry of Transport, Innovation and Technology (BMVIT), which is now the Federal Ministry for Climate Action, Environment, Energy, Mobility, Innovation and Technology (BMK). According to the country's strategy for the promotion of alternative propulsion systems and fuels as well as the Austrian Action Programme on Automated Mobility 2019–2022 [15], it is expected that vehicle fleets will be both automated and electric in the future since electric vehicles are more reliable due to less moving parts compared to an internal combustion engine car, while autonomous vehicles need the electrical brainpower to manage the perception, guidance and decision making tasks. Furthermore, an overall optimisation of the mobility system is expected. An example of a project funded under this scheme is DigiBus Austria, which does research and demonstrations of automated and electric shuttles in an intermodal regional mobility system. One of the synergetic features explored by the project is the precise positioning of the self-driving shuttle at bus stops by the well-controllable electric motors at low speeds.

5 International Benchmarks

The analysis of innovation strategies and programmes in the international domain beyond Europe shows that many countries are putting a strategic emphasis on fostering both the automation and the electrification of road mobility. However, the opportunities and challenges of thinking automation and electrification together have been addressed to much different extend so far:

The United States of America intend to become a world leader in both automation and electrification as it has clearly been stated by the Federal Government [16, 17]. An integrated view of these paths had been expressed by the second to last administration only [18]. Recently, the Federal State of California has clearly announced that all autonomous cars would be required to be zero-emission by 2025 [19]. At the same time, pilot providers of self-driving robotaxi services like Waymo and Zoox are already using electric vehicles to a large extend, as do the many trials with automated shuttles.

China is strongly fostering electric mobility. By 2035, the road vehicle fleet shall be half electric and half hybrid. Also, highly automated and connected driving shall be scaled by then [20]. Synergies between automation and electrification are not particularly pushed by the state, but are occasionally covered in test fields, e.g. Beijing E-Town or Shanghai International Automobile City.

Japan is a strong promotor of electric mobility as well. By 2035, the road vehicle fleet shall be 100% electric [21]. In parallel to this, automated driving is developed as part of cross-ministerial Strategic Innovation Promotion Program (SIP) [22] based on intelligent transport systems (ITS) infrastructures and services. The synergies and complementarities of automation and electrification have hardly been considered so far.

South Korea aims to become a world leader in automated vehicle technology by 2030. At the same time, it wants to increase the share of battery electric and fuel cell vehicles by 33% until then [23]. Koreas most ambitious smart mobility project Urban Connected Automated Shuttle Systems aims to deploy more than 200 automated and electric mini buses in Sejong, the country's new administrative capital.

6 Conclusions and Outlook

As pointed out in this chapter, there are numerous synergy potentials and complementary effects at both the enabling technologies and the systems levels if the automation and the electrification are not considered as separate or just coincidence, but mutually interacting, even merging technology paths. However, with the exception of some cases in the European Union and its Member States, these potentials have hardly be considered in the industrial and academic innovation strategies and public funding policies so far. Supposedly, due to the growing awareness of the systemic nature of both innovation paths and due to obvious similarities in the enabling technologies, this will change in the future.

The complete hand-over of the driving task from a human to a machine in fully autonomous vehicles requires the systems for environment perception, decision making and control to meet highest safety and performance standards that is still out of reach. At the same time, the imminent shortage of fossil fuel supply and the legislation in terms of CO_2 and tailpipe emissions demands a radical shift towards electricity and hydrogen as energy carriers for automobiles, which can hardly be imagined without a massive buildup of charging/filling infrastructures, further increase of battery capacities and a significant reduction of energy consumption.

On the longer timescale, and for both safety and energy, these trends will imply a shift from a bottom-up control logic based on individual vehicles to a merely top-down systemic control paradigm, requiring data and energy flows, software updates, and hardware allocation to be flexibly aligned at vehicle, infrastructure and cloud levels. Eventually, this will pave the path for road transport to enter into a new quality of merger between automation and electrification that strengthens sustainability and resilience in a comprehensive way.

Acknowledgements. This chapter summarizes and concludes the activities of the working group Task 29 "Electrified, Connected and Automated Vehicles" of the Hybrid and Electric Vehicles

Technology Programme (HEV-TCP) of the International Energy Agency (IEA), which carried out analysis and research work, organized expert workshops gathered stakeholders from academia, industry, and public authorities at different occasions in recent years. Gereon Meyer as the Operating Agent of Task 29 is grateful to the members of the HEV-TCP for lively discussion and helpful inputs, and to Germany, Austria and the United Stated of America for their continuous formal support of the Task. The authors would also like to express their gratitude to European Commission for funding this work through the Coordination and Support Actions SCOUT, COSMOS and particularly FUTURE-HORIZON.

References

1. Global Electric Vehicle Outlook, IEA, Paris (2022)
2. Taxonomy and definitions for terms related to driving automation systems for on-road motor vehicles (J3016_202104). SAE International (2021)
3. Meyer, G.: Synergies of connectivity, automation and electrification of road vehicles. In: Meyer, G., Beiker, S. (eds.) Road Vehicle Automation 3. LNMOB, pp. 187–191. Springer, Cham (2016). https://doi.org/10.1007/978-3-319-40503-2_14
4. Positionspapier Schlüsselrolle der E/E-Architektur und der Bordnetze für das Automobil der Zukunft. eNOVA Strategiekreis Automobile Zukunft, Berlin (2020)
5. Stevens, T., et al.: Assessing energy impacts of connected and automated vehicles at the U.S. national level—preliminary bounds and proposed methods. In: Meyer, G., Beiker, S. (eds.) Road Vehicle Automation 5. LNMOB. Springer, Cham (2018). https://doi.org/10.1007/978-3-319-94896-6_10
6. Roadmap für die Automobilität der Zukunft: Grundstein einer Innovationspartnerschaft. Forschungsvereinigung Automobiltechnik, Berlin (2021)
7. European Commission: STRIA Roadmap on Connected and Automated Transport: Road, Rail and Waterborne. European Commission, Brussels (2019)
8. European Commission: Electrification of the Transport System: Expert Group Report. European Commission, Brussels (2017)
9. Strategic Research and Innovation Agenda 2021–2027: European leadership in safe and sustainable road transport through automation. CCAM Partnership, Brussels (2022)
10. Strategic Research and Innovation Agenda 2021–2027. 2Zero Partnership, Brussels (2022)
11. Strategic Research and Innovation Agenda 2022: Electronic Components and Systems. KDT Partnership, Brussels (2022)
12. Aktionsplan Forschung für Autonomes Fahren. Die Bundesregierung, Berlin (2019)
13. FTI-Strategie Mobilität: Innovationen in und aus Österreich für ein klimaneutrales Mobilitätssystem in Europa. Bundesministerium für Klimaschutz, Umwelt, Energie, Mobilität, Innovation und Technologie, Vienna (2020)
14. Mikroelektronik aus Deutschland – Innovationstreiber der Digitalisierung. Rahmenprogramm der Bundesregierung für Forschung und Innovation 2016–2020. Bundesministerium für Bildung und Forschung, Berlin (2018)
15. Austrian Action Programme Automated Mobility 2019–2022. Bundesministerium für Verkehr, Innovation und Technology, Vienna (2018)
16. Ensuring American Leadership in Automated Vehicle Technologies: Automated Vehicles 4.0. National Science and Technology Council and U.S. Department of Transportation, Washington, D.C. (2020)
17. Executive Order on Strengthening American Leadership in Clean Cars and Trucks. The White House, Washington, D.C. (2021)

18. The Transforming Mobility Ecosystem: Enabling an Energy-Efficient Future. U.S. Department of Energy, Washington, D.C. (2016)
19. Senate Bill 500: Autonomous Vehicles: Zero Emissions. State of California, Sacramento (2021)
20. New Energy Vehicle Industrial Development Plan for 2021 to 2035. State Council of the People's Republic of China, Beijing (2020)
21. Green Growth Strategy through Achieving Carbon Neutrality in 2050. Ministry of Economy, Trade and Industry, Tokyo (2020)
22. Kuzumaki, S.: The Challenges for Automated Driving Systems Realization in Japan; SIP-adus. In: Meyer, G., Beiker, S. (eds.) Road Vehicle Automation 8. AVS 2020 2020. LNMOB. Springer, Cham (2021). https://doi.org/10.1007/978-3-030-80063-5_2
23. Future Car Development Strategy 2030. Ministry of Trade, Industry and Energy, Seoul (2019)

Part II: Business Models and Operations

Automated Vehicle Fuel Efficiency Town Hall

Dimitris Assanis[1(✉)], Avi Chaim Mersky[2(✉)], Jihun Han[3], Therese Langer[2], Jeffrey Lidicker[4], Rick Mihelic[5], Gabor Orosz[6], and Marina Sofos[7]

[1] Stony Brook University, 131 Light Engineering, 100 Nicholls Road, Stony Brook, NY 11790, USA
dimitris.assanis@stonybrook.edu
[2] American Council for an Energy Efficient Economy, 529 14th Street NW, Ste. 600, Washington, DC 20045, USA
{amersky,tlanger}@aceee.org
[3] Argonne National Laboratory, 9700 S. Cass Avenue, Lemont, IL 60439, USA
jihun.han@anl.gov
[4] California Air Resources Board, 1001 I Street, Sacramento, CA 95814, USA
jeffrey.lidicker@arb.ca.gov
[5] North American Council for Freight Efficiency, 6041 Stellhorn Road, Unit 15935, Fort Wayne, IN 46885, USA
mihelic2@verizon.net
[6] University of Michigan, G034 AL, 1231 Beal, Ann Arbor, MI 48109-2133, USA
orosz@umich.edu
[7] Advanced Research Projects Agency – Energy, U.S. Department of Energy, 1000 Independence Avenue SW, Washington, DC 20585, USA
marina.sofos@hq.doe.gov

Abstract. AV and CAV technologies are already affecting on-road energy usage and in the future may drastically change vehicle energy usage and efficiency. The Efficiency Town Hall, at ARTS2021 showed the importance of the question of how to balance individual vehicle efficiency with systemic transportation efficiency as well traffic and demand management. The Town Hall also showed that these questions can no longer be considered a problem for the future, solutions must be found for the vehicles entering the market now. Connectivity will also be key in ensuring that AV technology delivers consistent energy reductions. Finally, regulations which can capture the effects of CAVs and incentivize energy efficiency must be promulgated to ensure that AVs are designed with efficiency and energy reduction in mind.

Keywords: Autonomous vehicles · Connected vehicles · Energy · Demand · Energy efficiency · Energy consumption · Energy policy

1 Introduction

Breakout sessions B301 and B307 at the Automated Road Transportation Symposium 2021 (ARTS2021) were part of a double feature sponsored by AMS30(3), the Transportation Research Board (TRB) Subcommittee on Energy and Demand Implications;

G. Meyer and S. Beiker (Eds.): ARTSymposium 2021, LNMOB, pp. 53–70, 2023.
https://doi.org/10.1007/978-3-031-11112-9_6

AMS30, the TRB standing committee on Transportation Energy and ACP30, the TRB standing committee on Vehicle-Highway Automation. The double session focused on understanding the technical and regulatory hurdles of implementing efficiency regulations for automated vehicles, a topic that has yet to be tackled in other symposia. Vehicle automation is often promoted to increase safety and convenience, however little discussion often surrounds the associated energy and environmental implications of this new technology class.

A robust representation of professionals from industry groups, government agencies and regulators, academics, national laboratories, and public interest groups convened to open the lines of communication between regulator, stakeholders, and researchers. Presenters delivered twelve (12) presentations of 15–20 min with two (2) 45 min moderated panel Q&A at the end of each session. The B301 morning session featured seven (7) presentations followed by a panel discussion that focused on energy demand analysis of automated vehicles and enabling technologies that promote energy efficiency. The B307 afternoon session featured five (5) presentations followed by a panel discussion focused on understanding the challenges of developing a regulatory framework to regulate automated vehicle energy efficiency.

The ultimate goal of the sessions was to expose technical and policy research, promote data sharing and develop forward guidance and research needs statements around the themes of (1) understand the true financial cost of implementing Automated vehicles (AVs) energy efficiency regulations – or the environmental costs associated with delaying and (2) exposing the "bleeding-edge" energy efficiency enabling technologies/research and analysis. The symposium outcomes described hereinafter regarding the importance of balancing individual vehicle fuel economy with system-wide energy use/reduction objectives will ultimate be monitored by the TRB subcommittee on Energy and Demand Implications AMS30(3) and future breakout sessions at the Automated Road Transportation will be tailored to ensure this discussion is continued and adapted to the changing landscape.

2 Summary of Presentations

The following section summarizes a selection of the research presented during the conference breakout session. Each presentation's section summarizes the findings and/or methods, where applicable, of one of the presentations. Presentations included original research, summaries of literature and syntheses reviews, reports from regulators and government agencies, and reports from industry groups. A list of all presentations, including those summarized in this chapter, and slides for each can be found in the conference proceedings.

2.1 Autonomous Vehicles and Off-Cycle Emission Credit Testing

Avi Chaim Mersky, American Council for an Energy Efficient Economy

2.1.1 The Growing Impact of Automated Vehicles

Automated vehicle (AV) technology continues to be developed and commercialized and are already widely deployed and encompass many existing driver-assistance and safety features. Over a quarter of all new vehicles delivered to U.S. dealers in Q1 2020 had some automated features, while the market share of Level 2 AVs has grown from at least 2% of all new vehicles to at least 10% in just 2018–2019 [1–3]. Research by ACEEE suggests even more advanced AVs will be significant components of the US vehicle fleet by 2035 [4]. Vehicle fuel efficiency and emissions are both highly sensitive to how the vehicle is controlled and, therefore, automation. AV efficiency is also highly variable. A review of recent literature by ACEEE showed that AV features, likely to be available on mass market vehicles, could reduce vehicle efficiency by as much as 14% or increase it as much as 52% [4].

2.1.2 The Need for Autonomous Vehicle Efficiency Regulations

The current standard light-duty vehicle fuel economy and emissions test procedures rely on testing fuel consumption and emissions for fixed velocity schedules on a dynamometer. These procedures cannot detect the fuel economy impacts of technologies, including AV technologies, that change how the vehicle responds to the environment around it. While there is a mechanism to recognize the benefits of technologies that are not detected under the test procedures: the process is labor intensive for both automakers and the regulatory agencies and the results are not guaranteed. Both factors act as a cost that reduces the incentive to improve AV efficiency. Hence it is desirable that emissions and fuel economy regulations incentivize manufacturers to design AV systems with fuel efficiency in mind.

2.1.3 How Autonomous Vehicle Fuel Efficiency Should Be Regulated Now

Our recommendations apply only to level 1–3 AVs. We believe that highly or fully autonomous vehicles must have their fuel efficiency be tested as a single unit, rather than applying credits to specific features. We propose that the regulating agencies, EPA and NHTSA, define discrete AV Feature Groups (AVFG) that describe a set of unique operating conditions and capabilities. Additionally, AVFGs should be separated by limits of certified, not effective, functionality, even if this leads to identical divisions of driver and computer control and responsibilities. Certification should be based upon manufacturer instructions, unless and until NHTSA starts issuing requirements for AV safety certification.

The regulating agencies should provide a list of AVFGs eligible for credits and develop standardized rules on how these AVFGs should be evaluated. The agencies should publish the rules for public comment. The final test protocols should include both the vehicle testing methods and specific rules on how these results will be used to calculate credits. These credits should be based on regularly updated estimates or regularly updated empirical evidence of the extent of technology use and, if significant, the technology's penetration rate.

Over the short term, these suggestions could potentially be implemented under the existing optional off-cycle credit program. Over the longer term, we believe that the agencies should consider requiring that all AVFGs be tested for fuel economy changes. The resulting changes, even if negative, should be applied to the vehicle's rated fuel economy on a mandatory basis rather than as an optional credit. This will ensure that applications that increase fuel consumption will be accounted for. AV efficiency improvements can also be considered by policymakers when setting efficiency standards.

2.1.4 Future Work

Our recommendations reflect on an existing regulatory environment that is concerned with the direct effects of technology on an individual vehicle's fuel economy. The impact of AVs is both dependent on surrounding traffic and also changes traffic conditions. These systemic impacts need to be better understood and the regulatory agencies should ensure that they do not encourage technologies whose systemic detriments are greater than their individual benefits. AVs may also change the total demand for travel. While existing efficiency and emission are not intended to tackle such impacts on energy use, regulators and policy makers should create policies to ensure that AVs do not increase total emissions, even if increase vehicle efficiency.

2.2 National Academies Light-Duty Fuel Economy Report: Findings on CAV Technology Energy Impacts

Therese Langer, American Council for an Energy Efficient Economy
The recent National Academies of Science Engineering and Medicine study *Assessment of Technologies for Improving Fuel Economy of Light-Duty Vehicles—Phase 3* (https://www.nap.edu/catalog/26092) examined vehicle efficiency technologies likely to be available in 2025–2035 [5]. The committee relied on information gathered from industry meetings and site visits, public information sessions, expertise of committee members, and the literature. The study, sponsored by U.S. DOT's National Highway Traffic Safety Administration, was mandated by Congress in Energy Independence and Security Act of 2007. [This presentation summarized the study's findings on the energy impacts of connected and automated vehicle (CAV) technologies; recommendations on associated policies were discussed in Sect. 2.4].

The study distinguished energy issues for lower level CAVs (SAE Levels 1–3) from those of fully autonomous (Levels 4 and 5) vehicles, focusing in the former case on effects of the technologies on the fuel economy of individual vehicles. For autonomous vehicles, there is a much wider array of potential energy effects, based on these vehicles' implications for car ownership decisions, mode choice, vehicle miles traveled, and other issues that go beyond the technology's effects on the vehicles themselves.

The study summarized cost and effectiveness of three CAV technology packages.[1] Key findings and caveats included that low levels of automation (Level 2) can provide fuel savings of up to 8% through optimizing velocity and minimizing acceleration events, though the savings depend strongly on driving conditions and powertrain type. Adding

[1] See Table 8.6 of the National Academies study.

connectivity to increase the system's prediction horizon and optimizing power train controls allows fuel savings of as much as 20%, with the greatest benefit achieved in plug-in hybrid vehicles on trips exceeding the battery range. These estimates do not represent savings on standard test cycles, however, nor do they reflect energy effects of any changes to traffic flow the CAVs may produce. All-electric vehicles will see the lowest efficiency gain but will benefit from other synergies with CAV technologies.

The committee estimated direct manufacturing costs of the Level 2 package at $1,520 and Level 2 with power train controls and connectivity at $2,410, with modest declines in the costs of both packages over the next 15 years. The fully autonomous vehicle (Level 4/5 with connectivity) was estimated at $7,210–$17,210, depending on lidar unit specifications, but was projected to decline to $2,545–$4,683 by 2035. Since CAV technology adoption is largely driven by benefits other than fuel savings (safety, mobility, convenience), these costs should not be attributed entirely to fuel savings in the context of a cost-effectiveness assessment of technologies for regulatory purposes.

For autonomous vehicles, the committee highlighted a recent national laboratory meta-analysis of the literature, which bounds likely energy impacts of full adoption of autonomous vehicle between a 40% reduction and a 70% increase in energy use [6]. While power draw of these higher level CAV systems can be substantial—on the order of 2 kW—the draw for a fixed vehicle capability will decline rapidly over time as electronic systems evolve. However, total electrical load of these systems may remain significant as their functionality increases, due especially to growing computing requirements.

The study found that connectivity is unlikely to be widely deployed in 2025 but could reach high adoption levels by 2035 if public infrastructure is updated to collect, process, and distribute data, and if useful, affordable connectivity services are available. Autonomous vehicles' share of the market in 2035 is likely to fall in the 0–40% range, with ride hailing and delivery fleets accounting for 40–60% of those sales.

2.3 National Academies Light-Duty Fuel Economy Report: Policies to Promote CAV Technology Energy Savings

Therese Langer, American Council for an Energy Efficient Economy

Among the recommendations of the recent National Academies of Science Engineering and Medicine study Assessment of Technologies for Improving Fuel Economy of Light-Duty Vehicles—Phase 3 were several on federal agency actions to promote energy savings from connected and automated vehicle (CAV) technologies. [For a brief description of the National Academies study, see Sect. 2.3 on energy impacts of CAVs.]

For lower levels of vehicle automation (SAE Levels 1–3), the study considered primarily effects of these technologies on fuel economy and greenhouse gas emissions of individual vehicles and implications for vehicle standards. The study found that CAV technologies enable, but do not ensure, substantial fuel efficiency improvement over current vehicle technologies. Today's vehicle test procedures generally cannot detect any CAV technology fuel efficiency benefits, so these technologies could only help manufacturers comply with fuel economy standards through the off-cycle credit program.

While off-cycle credits could promote adoption of CAV technologies, the agencies will need to exercise caution in awarding such credits due to the complexities of evaluating CAV energy impacts. In particular the committee recommended that 1) off-cycle credits be available for CAV technologies only to the extent they improve the fuel efficiency of the vehicle on which they are installed (and not through changes in traffic flow, for example), and 2) any credits be based on realistic assumptions regarding technology adoption on other vehicles or infrastructure. Moreover, given that some CAV technologies are becoming commonplace, once their energy impacts have been adequately quantified the agencies should consider their potential benefits in setting the level of the standards.

With regard to quantifying CAV technology impacts for purposes of compliance with vehicle standards, the committee noted that allowing these vehicles limited departures from the standard cycles during testing would permit some CAV technologies' fuel efficiency gains—and losses—to be measured. More generally, the committee found the problem of estimating CAV technology energy impacts to be symptomatic of a larger issue in the fuel economy standards program, namely the divergence between vehicles' fuel economy as captured in testing and their performance in the real world. The study underscored the opportunity and need to rely more on real-world data to assess vehicles' performance, noting that "vehicles currently being produced/sold in the U.S. market can record fuel consumption over specific periods of time, which provides the capabilities for verifying performance and could enable a shift from the test-cycle-based approach of estimating emissions to an approach of directly measuring emissions."

In the case of fully autonomous vehicles, the study noted that the maximum feasible fuel economy standards for these vehicles in fleet use could be more stringent than standards for personally owned vehicles, and that an all-electric mandate should be considered for autonomous fleet vehicles. However, achieving positive energy outcomes through adoption of autonomous vehicles will require a much more extensive policy approach. Agencies should consider actions to guide system effects of autonomous driving, including policies to promote vehicle sharing and ensure these vehicles' complementarity to less energy-intensive modes. Additional research and policies are needed to advance the simultaneous achievement of the safety, economic, environmental, and equity benefits which autonomous vehicles can provide.

2.4 Impact of Vehicle Automation on Energy Consumption[2]

Jihun Han, Dominik Karbowski, Jongryeol Jeong, Namdoo Kim, Julien Grave, Daliang Shen, Yaozhong Zhang, Aymeric Rousseau, Argonne National Laboratory

Connectivity and automation technologies offer the potential for improving vehicle efficiency through energy-focused controls. Under the SMART 1.0 (Systems and Modeling for Accelerated Research in Transportation) Mobility Laboratory Consortium [6], we developed various automated driving controllers, e.g., "speed-only" optimization [7], "speed + powertrain" co-optimization [8]. The speed + powertrain algorithm co-optimizes speed and powertrain to achieve maximum efficiency. Using RoadRunner, a new simulation framework for research energy-efficiency and driving

[2] This material is based upon work supported by the U.S. Department of Energy, Vehicle Technologies Office, under the Systems and Modeling for Accelerated Research in Transportation (SMART) Mobility Laboratory Consortium, an initiative of the Energy Efficient Mobility Systems Program.

automation, we performed a large-scale simulation study applying the algorithms and demonstrated up to 22% savings when utilizing traffic signal information through V2I (Vehicle-to-Infrastructure) communication.

However, the algorithms need to be deployable on real-time control units and provide the energy savings on real vehicles without safety issues (e.g., traffic rule violations, rear-end collisions). To this end, we have developed an XIL (anything-in-the-loop) workflow that includes: creating a digital twin of a real vehicle and environment, developing an automatic building process from full simulation to a mix of simulation and hardware, developing a methodology for interactions between a real vehicle and a simulated environment, developing an automatic quality-check process for control functionality verification, etc. The XIL workflow accelerates the experimental testing process, enables testing of various control algorithms and quantification of their impact on energy consumption, while ensuring high test-to-test repeatability and accuracy.

We improved the speed-only optimization algorithm to perform well in a broad range of situations using RoadRunner, and implemented it in the real vehicle to automatically drive in an energy-efficient way. Finally, we tested the automated driving controller for 22 scenarios (total 280 km) and applied it to two powertrains (GM Electric Bolt and Blazer). Scenarios defined by a combination of route and controller features (e.g., V2I communication on/off, preceding vehicle speed prediction on/off) include various situations such as traffic light approach, speed limit change, and traffic. The controlled ANL on-dynamometer tests validated all functionality and performance of the automated driving controller and led to a successful on-track (3.72 km) demonstration at ACM (American Center for Mobility) [9]. Experimental test results showed that energy savings from V2I communication become greater (up to 30%) for single intersection approach and departure situation on empty road, as the remaining time to the next green light is longer. In scenarios with traffic, energy savings are increased (about 11%) as the penetration rate of V2I communication increases (0%, 50%, and 100% in 2 vehicle scenarios). A vehicle without V2I following the virtual preceding vehicle equipped with V2I communication also saves energy (about 10%). Moreover, more accurate and longer prediction of the preceding vehicle's driving behavior (e.g., braking-stop-wait-departure at a red traffic light) can generate smoother trajectories and more energy savings (about 7%). Note that these energy saving values are computed with respect to the controller without V2I communication (not a human driven vehicle).

In future works, we would like to validate energy impacts for real-world representative scenarios designed well by data. Moreover, we could test advanced controllers (e.g., enabling multi-traffic light approach, speed + powertrain control) to gauge their further energy saving potentials through ANL xIL workflow.

2.5 Automated Vehicle Policies for Equity and Clean Air

Jeffrey Lidicker, California Air Resource Board

2.5.1 Background

The State of California has been actively tracking and researching Automated Vehicles (AVs) as initial information indicates that AVs may influence emissions dramatically. A recent study by Dr. Merksy indicates that AVs can, depending on how well they are programmed to eco-drive, reduce vehicle emissions by as much as 40%, or increase them by up to 14% [4]. A study by Dr. Hardman et al. indicates that 36% of drivers using available partial automation features reported "more long distance travel" and 40% reported "more driving during periods of congestion" [10]. The study estimates that, for Teslas only, due to partial automation an average of 4,884 additional miles are driven per year per vehicle. Lastly, a study by Dr. Wadud et al. estimates that energy consumption and emissions from AVs could be cut in half, or double depending on the particulars of how they are operated [11]. Certainly, if AVs were to double energy consumption or emissions, this would derail California's emissions reduction goals [12].

In 2018, a multi-agency workgroup was formed in California to ask these policy questions. Over 10 state agencies participated in the workgroup.[3] The workgroup produced an AV Principles for Healthy and Sustainable Communities document [13]. Although the document was adopted by the Governor and subsequently posted to the Office of Planning and Research website, it does not officially represent the position of the participating agencies or commissions. It exists, however, for policy makers from local, state, and federal agencies to use as a resource. The AV policy document lists eight guiding principles.

2.5.2 AVs as Shared-Use Vehicles

With respect to energy consumption and therefore emissions, it is preferable for AVs to be shared-use vehicles instead of privately owned. If AVs enable a high percentage of shared trips, say 85%, then there would be essentially a de-facto VMT fee in place without any new legislation, new authority, or government run administration and reporting system. These shared ride fees are based on a combination of time and distance along with a built-in peak pricing mechanism, which would be an optimal VMT reduction policy. Other attributes of this policy are better utilization of vehicle capital and reduced parking demand that enables better utilization of high-value real estate.

2.5.3 AV Rides as Pooled Rides

Maximizing the average number of passengers in an AV will reduce vehicle miles traveled but not passenger miles traveled. The higher the penetration of shared-use vehicles, the

[3] Participating agencies included but are not limited to: CalEPA, CalSTA, Caltrans, CARB, CDPH, CEC, DGS, DMV, Go-Biz, OPR, and SGC. Also participating was the CPUC.

more opportunities for pooled rides. Other benefits of pooling include lower prices that improve transportation equity and fewer empty miles traveled. For example, a policy designed to increase the use of pooling is the California Clean Miles Standard, in which shared-ride services must meet grams of CO_2 per passenger mile traveled targets [14]. Thus, the more they pool, which reduces VMT but not PMT, the easier it will be to meet the targets. Companies that have developed and continue to develop pooling services are Via, Lyft, and Uber.

2.5.4 AV's as Low-Emission Vehicle

Any policy that can motivate AVs to be low-emission or zero emissions will produce fewer emissions than one that runs on fossil fuels. For example, the proposed California Senate Bill 500 would require all light-duty AVs be zero-emission by 2031 [15].

2.5.5 Right-Sized AVs

Rightsizing is a term that implies that, on average, the number of available seats in a car is equal to the number of passengers on a particular trip. Thus, if a city has 85% of trips by single-passenger travelers, then 85% of the vehicles used for trips would be one-seaters, and so on. Policies that discourage driving with empty seats will reduce emissions overall. Rightsizing might even reduce congestion as four single AVs may fit in the same space as a large SUV at a red light. In 2015, Dr. Greenblatt at LBNL estimated that vehicle rightsizing could reduce energy consumption, and therefore, emissions as much as 45% [16]. AVs would be necessary to achieve optimal rightsizing.

2.5.6 Integrate AVs into Multimodal Systems

Imagine if all AV policy was dictated only by profit and AVs had no bicycle racks on them, were programmed to pass bicycles very closely, and were not allowed to take anyone to or from a transit station or let anyone out of the vehicle when stuck in gridlock. These policies might improve profits for ride providers but would likely discourage the use of multi-model transport systems such as trips that make use of more than one mode: bicycles, transit, walking, and AVs. Instead, imagine AV policies that encourages the use of transit and trips with more than one mode. Examples of such policy are the CA Clean Miles Standard regulation that offers compliance credits for ride hailing companies to integrate transit into their mobile applications [14], and the company Via that has been partnering with transit agencies to provide on-demand transit in settings where fixed route transit isn't providing good access [17].

2.5.7 Shared AVs in Planning Polices

The sixth and seventh guiding principles are closely related. The sixth one encourages land-use policies that leverage shared AVs in ways that encourage infill rather than sprawl. For example, cities can leverage shared AVs to reduce the need for parking requirements freeing up land for a myriad of other uses such as housing or greenspace. Reducing parking requirements for buildings would also reduce housing costs and availability, which could improve equity metrics in non-transportation ways.

The seventh guiding principle applies to complete and livable streets - the spaces between the land or city blocks. Policies that can leverage shared AVs to prioritize people, other modes, and overall health and safety. For example, AV policies can motivate AVs to be polite to pedestrians and bicyclists making streets safer for pedestrians and other forms of active transportation. Other policies can allocate curb space for shared AV ride or freight drop-off and pick-up, so that loading and unloading passengers and freight do not block traffic lanes reducing traffic congestion. Due to advantages of AVs, perhaps only two traffic lanes are needed instead of three so that more street space can be used for people, other modes, and beautification. Parklettes are another example of a benefit allowed by the combination of city policy and the reduced demand for parking afforded by AVs. Parklettes provide a higher quality of life and increased revenue for restaurants among other benefits.

2.5.8 Transportation Equity and AVs

AVs present an opportunity to improve equity in transportation by increasing access or mobility with lowed transportation costs. Several features of AVs can lower operating costs such as removing the cost of a driver and spreading fees across multiple passengers when pooling rides. Capital costs can also be lowered due to higher utilization of shared vehicles for a lower cost per mile. Together, these two types of AV cost reductions improve the feasibility of on-demand transit that can expand service availability into disadvantaged communities and offer the potential of mobility for the disabled and elderly at the same price as for anyone else. However, without AV policies that ensure private companies provide services to everyone everywhere, and not just where the highest profit margins are, the opposite could happen.

2.5.9 Conclusion

The potential for AVs to improve transportation access for all, including for disadvantaged or disabled and elderly communities, is unprecedented. AVs can also improve health and safety by reducing accidents, reducing vehicle emissions, and encouraging active transportation. However, without government policies, these improvements may not happen.

2.6 Energy Efficiencies of Trucking Automation Now and into the Future

Rick Mihelic, North American Council for Freight Efficiency
The conversation on automation starts by understanding that automation is a journey, not a destination. Continuous improvement is the nature of heavy truck technologies. Freight efficiency is about moving more freight with less energy *and less cost*. Cost and energy are intertwined. Commercial trucks are businesses. They need to be profitable.

The future is easy to predict, after it has happened. History is replete with technology marvels that did not fare so well in the market. We are in the midst rapid and diverse changes in trucking technology. Zero and net zero solutions are ramping up. All have infrastructure needs. Automated and connected vehicle technology is just part of the story. And digital data and data mining is coming with all of these. This transition is being driven both by market and regulatory forces.

Nearly every day we see new AV companies and products in the news. These vehicles are on the road in limited numbers already, and more are coming. Why are AVs coming? There are multiple factors that can be grouped under the headings people, market, and accidents. Human employees bring with them a wide range of overhead factors. The market demands moving more freight than we have drivers. Competition requires all companies to minimize costs while maximizing profits. The convergence of all these factors, shown in Fig. 1, is the opportunity for AVs.

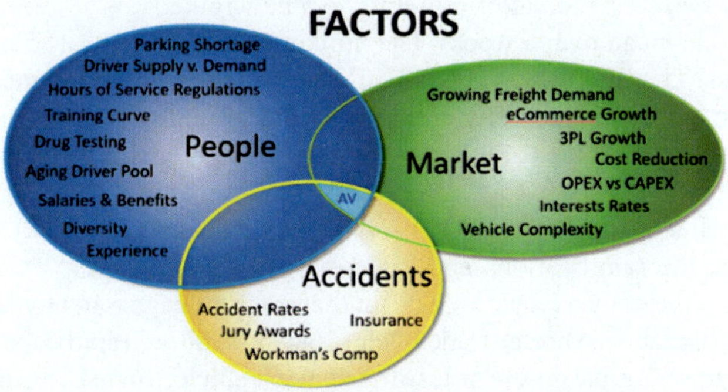

Fig. 1. Factors driving AV development and adoption

Every technology has tradeoffs, advantages, disadvantages, and unknowns. While AVs lower operating costs, they increase capital costs. AVs can increase daily volume of freight moved, but 24/7 operation brings with it a range of infrastructure challenges. While AVs promise to lower the number of accidents, the severity of the accidents that do occur could be more severe. These are just some examples.

The big question is always how much improvement? The answer is very context sensitive. It depends. Physical testing and analytical models range in real world performance from negative improvement to 0%, 5%, 10%, 20% and more. NACFE has quoted research showing that technology makers tend to over-estimate their products capability by as much as 3 times. While consumers of that technology tend to underestimate that same performance by as much as 3 times. Reality is usually somewhere in between. Not as good as the manufacturer's vision, not as bad as the fleet's expectations. In the end, both want the technology to work well. But there are no average fleets, no average drivers, no average trucks, no average routes, no average loads. Your savings may differ. There are no SAE standards yet for evaluating fuel economy in traffic conditions.

The savings also need to be in context of the entire freight system. A holistic view is the Total Cost of Ownership, or TCO perspective. It is common to look at operating costs on only a per truck basis. In stable periods this is about 1/3 of operating cost is due to the equipment, 1/3 due to the driver and 1/3 due to fuel. One argument for automated trucks reducing operating costs. But there is more to that.

Trucking has always needed more people, across the board. It is not just drivers. Its technicians, back-office people, supervisors. Competition for workers has grown, and quality of life factors are weighing more on job choices. Automated vehicles present an opportunity to add freight capacity.

Many factors are contributing to the shortfall in trucking people. In talking to fleets, in many cases it is not a lack of applicants. It is a lack of "qualified" applicants. The emphasis is on experienced, skilled drivers. Automated trucks are expected to fill this void.

So how do AVs help solve freight issues? Look at a 24-h day example. On an actual one-day truck route a particularly good driver and truck achieved 10.6 mpg and on a 637-mile route in an 11 h driving period. An automated truck does not need to stop for breaks, this gives it some advantage also in net fuel economy, allowing it to arrive earlier. It is then free to be reassigned after refueling to a new route.

What does this mean over a week? The human driver can get 3,185 miles with five similar deliveries. The automated driver in this case can do 8,918 miles and 14 deliveries. This is about three times the capacity of the single driver. This is just one example. There are a lot of duty cycles, routes and trips. Some drivers go back and forth A to B to A. Some have multiple stops A-B-C-A. Others may rarely get home, picking up new loads and routes at each stop, an A-B-C-D-E-F-. And the distances vary a lot.

So, what are the ramifications and trade-offs. An autonomous vehicle may make nearly three times the deliveries per week. But that extra mileage is not free. It brings with it increased maintenance. Shorter trade cycles. Need for more rapid capital. Regarding accident rates, they are based on miles driven per vehicle. Increasing the miles by a factor of three increases the opportunity for accidents to occur, while the technology is working to try to reduce the accident risk. The delivery network also must be able to accommodate 24/7 operations. And software is not free.

Automation is also in context of parallel movement towards zero and near zero emission vehicles. Some are competing for money and resources, some are enabling AVs.

So, what if the driver is not in the vehicle? Where does truck design lead? Moving more freight may be possible with larger trailers. While staying inside today's legal lengths. And what if we combine a number of technologies? We could see automated road trains.

AVs have the potential to help move more freight efficiently. How much depends on a holistic view of the freight system. For the near term, they will supplement not replace traditional vehicles. There are tradeoffs and unknowns with all new technologies. The market will prove out AV technology over time.

2.7 Infrastructure Assisted Automated Driving on Highways

Gábor Orosz, University of Michigan

Connected road infrastructure (CRI) can dramatically improve transportation system-level energy efficiency, productivity, and emission by exploiting connected and automated vehicle (CAV) technologies. This is expected to lead to significant improvement in the efficiency of passenger and freight transport (measured in mile/hour/kWh) even for low penetrations of CAVs. As illustrated in Fig. 2, achieving such high efficiency of road transportation relies on the tight integration of connected automated vehicle

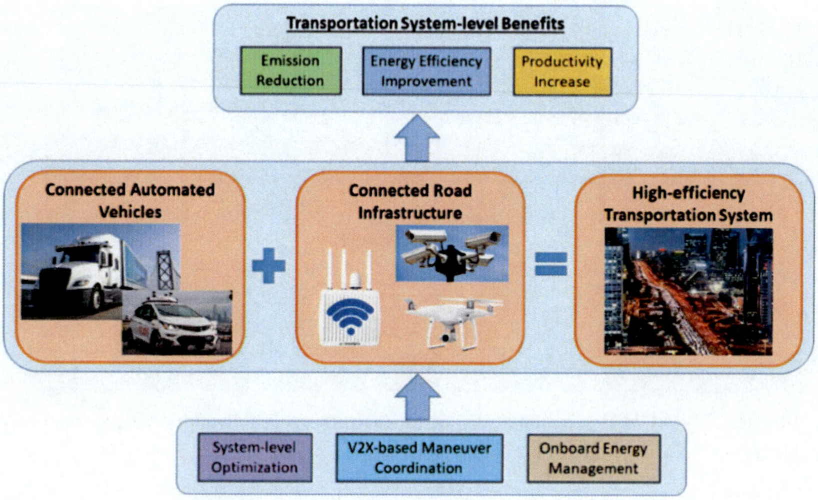

Fig. 2. Information flow of the integrated system

(CAV) technologies with connected road infrastructure (CRI) technologies. In particular, information collected via fixed-base and airborne cameras and vehicle-to-everything (V2X) communication shall be processed and aggregated by the CRI in order to merge the benefits having basic information about all non-connected vehicles and in-depth information about connected vehicles. Such information can enable CAVs to perform infrastructure-assisted automated driving: they can move through traffic faster and while using less energy delivering goods and passengers in a highly reliable manner. These actions can also be integrated with the onboard energy management systems of the CAVs in order to maximize energy efficiency at the component level. Since these CAVs also heavily influence the rest of the traffic they can lead to dramatic improvements of transportation system-level energy efficiency and significant reduction of emission even for low CAV penetrations. The arising highly efficient transportation system shall allow unprecedented growth of productivity with small investment to the infrastructure.

In case of a CAV, the efficiency improvements arise mainly from having access to lane specific real time traffic predictions for the next few miles ahead via V2X communication. These improvements are shown to be significant compared to the baseline scenario of having purely sensor-based automation [18–20]. Such strategies rely on technologies that make such information available for CAVs and on algorithms that allow these vehicles to achieve such improvements with high reliability.

A section of highway I-275 near Ann Arbor, MI is illustrated Fig. 3 where our team is currently deploying elements of the proposed infrastructure in collaboration with the Michigan Department of Transportation. This infrastructure will provide us with an unprecedented opportunity to monitor and predict road traffic. Historically, traffic data has been collected using fixed-base cameras mounted on roadside columns, and a set of those are already available along I-275. We are augmenting these with airborne cameras and with V2X communication devices that communicate to each other via 5G communication. This infrastructure enables us to collect high precision trajectory data from

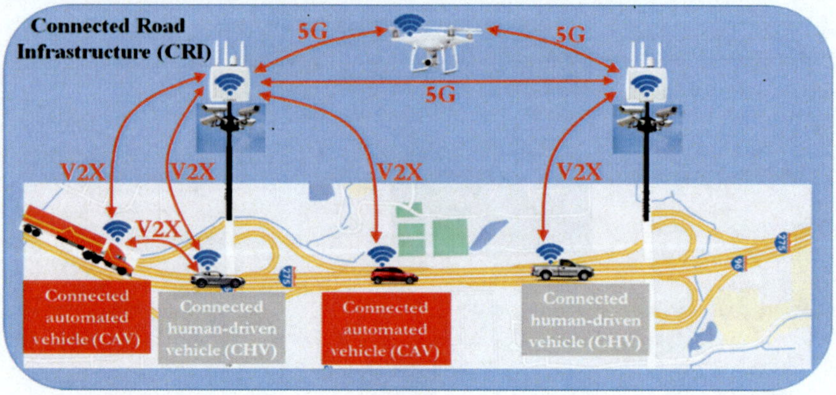

Fig. 3. Physical layout of the connected road infrastructure (CRI) supporting connected automated vehicles (CAVs) on highway I-275

passing human-driven vehicles, some of whom are equipped with V2X communication devices making them connected human-driven vehicles (CHVs). To minimize latency camera data will be fused with V2X data using edge computing on road-side units provided by Commsignia. As even the trajectory of a single CHV can provide prediction, this methodology will be able to accommodate different penetration levels of connectivity. The established CRI is able to communicate lane specific real time traffic predictions to the passing CAVs via V2X, enabling CAVs to select their lanes and longitudinal speed in order to maximize their efficiency.

Real-world traffic data collected on highway I-275 can also be used offline to design the connectivity-enhanced controllers while utilizing high fidelity vehicle models. This allows us to optimize the longitudinal controllers (engine, transmission, and brake) as well as the lane selection algorithms before implementing them on real hardware. Following such virtual development, we will utilize a Navistar class-8 connected automated truck, developed within DOE's Supertruck program, which is equipped with a real time controller giving access to the states of the engine, transmission, brakes, etc. Integrating the real time controller with a V2X onboard unit, we will make the truck capable of utilizing traffic predictions by the V2X road-side units at the Navistar Proving Ground. This will allow us to test the proposed algorithms in a safe environment. Finally, the truck will be tested in the real world on highway I-275 utilizing the real time lane specific traffic predictions provided by the deployed CRI. The developed technologies will be extended to trucks with higher level of automation with the help of Plus.ai and to different vehicle classes with help of General Motors enabling the team to evaluate the efficiency improvements of CAVs across a variety of vehicles with different drive types.

2.8 Improving the Energy Efficiency of Connected and Automated Vehicles: Results from ARPA-E's NEXTCAR Program

Marina Sofos, Department of Energy ARPA-E
The U.S. Department of Energy's Advanced Research Project Agency–Energy (ARPA-E) developed and initiated the NEXT-Generation Energy Technologies for Connected

and Automated on-Road Vehicles (NEXTCAR) Program in 2016 with the aim of utilizing connectivity and SAE L1-L3 vehicle automation to achieve a 20% savings in the energy consumption of conventional and hybrid electric cars and trucks. Under the NEXTCAR Program, eleven individually awarded project teams, in collaboration with 13 OEMs, suppliers and partners, developed and implemented new advanced vehicle dynamic and powertrain control (VD&PT) technologies utilizing 2016–2017 L0 baseline vehicles.

The first part of this talk included an overview of the achievements of the NEXTCAR Program for each of the technologies developed and evaluated on light-duty and medium-duty vehicle applications (a sub-set of 9 projects). Adding functionality to existing advanced driver-assistance systems (ADAS) integrated into L1-L3 vehicles allowed for readily attainable energy efficiency improvements of 20% for a range of vehicle propulsion technologies, including internal combustion engine vehicles (ICEs), hybrid electric vehicles (HEVs), and plug-in hybrid electric vehicles (PHEVs). Furthermore, it was shown that real-time powertrain optimization is facilitated by information obtained via connectivity across a range of time-scales. Results also showed a strong trade-off between elapsed trip-time and energy expenditure for typical vehicle operation. Finally, the power consumption of the sensing and computational systems required for connected and automated vehicle operation constitutes a major parasitic load (i.e. more than 1 kWe for each vehicle under the NEXTCAR Program) that needs to be considered in future vehicle designs.

The second part of this talk gave a preview of the second phase of the NEXTCAR Program that launched in 2021. Phase II builds on the goals of the original Program with a specific focus on light-duty passenger vehicles, a 30% reduction in energy consumption, and taking vehicles to Level 4 of automation, where a vehicle is able to perform all driving operations on its own with optional human override. The overall objective being to develop technologies (including those developed under the original Program) that will address potential runaway energy usage caused by higher levels of automation.

3 Conclusions

Automation technologies have long been promoted for their benefits in terms of improving safety and convenience for the end-user. However, it is also starting to become well understood that automation can decrease the barrier to mobility and the increased accessibility may be followed by an increase in travel demand which will yield higher annual vehicles miles traveled (VMT). Therefore, it is imperative to consider the energy efficiency of automated vehicles in order to counteract the potential increase in VMT they may cause.

The first overarching conclusion drawn from the presentations and the open panel discussions relates to need for policy and regulations to be developed and adopted to ensure that CAVs are designed for energy efficiency. Existing efficiency and emission regulations are not sufficient to capture the effects of CAVs and CAVs are not guaranteed to reduce emissions per VMT without conscience automaker design choices. Additionally, for the reasons stated above, increased in VMT are expected with increasing penetration of automation technologies and without intervention, there is little reason for automotive manufacturers to ensure energy efficiency of vehicles remains as high as possible or that increases to VMT do not lead to negative externalities in excess of any benefits from automation.

The second overarching conclusion drawn from the presentations and the open panel discussions relates to balancing the individual vehicle-level efficiency again system level efficiency improvements which is more desirable. An individual vehicle can achieve a hyper-localized maximum energy-efficiency. Correspondingly, the deployment of several of these highly efficient AVs should increase the energy efficiency of the overall system network, given all other variables remain constant. The reality though can be much different, given that AVs may not act like human-driven vehicles [21, 22] and as such their driving behavior may cause localized increase in traffic congestion. Thus, while the individual vehicle level efficiency is improved, the overall transportation system level energy efficiency can be degraded. Balancing, or potentially sacrificing individual vehicle-level energy efficiency may ultimately achieve a higher total system-level energy efficiency if traffic congestion can be mitigated.

The third overarching conclusion drawn from the presentations and the open panel discussions relates to balancing vehicle energy-efficiency and safety. Vehicle automation technologies have come a long way to improve safety and it is conceivable that in the far future, a significant portion of physical present-day vehicle safety requirements will be "virtualized" or "internalized" in the deepest layers of the operating logic of AVs. This will allow vehicles to be optimized for light-weighting and aerodynamic efficiency which will in-turn improve the vehicle energy-efficiency. For the near future, there will always be the question of when will safety in AVs reach the point to where we can downsize or eliminate certain crash-safety features (heavy sub-frames, large crumple zones, high should er lines, etc.) An additional point of consideration that requires striking the same balance and falls outside the vehicle relates to the safe following distance of two vehicles. In the case of co-operative and adaptive cruise control (CACC), decreased headway time can decrease the aerodynamic drag however, it can also significantly affect vehicle safety as the required stopping distance is violated. Ultimately, automation technologies will enhance safety and due consideration must be given as to how, where and if, vehicle safety is compromised in the name of vehicle energy-efficiency.

4 Next Steps

The Efficiency Town Hall showed the need for both further research and for new regulatory actions. Significantly more research is needed into how highly automated (L4/5) vehicles will affect traffic patterns and travel demand. Research is also need into what parts of automation (decreased cost of travel time, or changes in ownership models) will lead to these changes, as the policy levers to mitigate undesirable results may differ, depending on underlying causes. More research is also needed into how AVs both will and could affect transit demand. Panelists agreed that decreased transit demand would be an undesirable outcome, but research suggests that this outcome is not guaranteed. More research is needed into AV and transit interactions.

While new regulations are necessary to ensure AVs are designed for energy efficiency, more work is necessary to determine what these regulations should be, how they would even test energy efficiency and how they would balance individual vehicle performance vs. systemic effects on traffic and energy consumption. Significantly more research is needed into AV fuel economy testing, real world driving patterns and AV effects on traffic, as well as how lessons learned from simulations and physical studies can be applied in a regulatory space.

Finally, both more research and policy discussions are needed into the subject of crashworthiness of highly AVs. Vehicle light weighting represents an enormous opportunity to reduce energy usage and vehicle cost, but can only be done when AVs are "safe enough". Little agreement exists on where this point is and whether it is achievable only in purely autonomous environments, or if it can be achieved in mixed, AV and human, road systems.

References

1. NAS: Assessment of Technologies for Improving Light-Duty Vehicle Fuel Economy 2025–2035. National Academies of Sciences, Engineering, and Medicine, Washington, DC (2021)
2. Low, J., Doshi, R., Dutt, I., Jones, C., Ryan, M.: Canalys: 10% of new cars in the US sold with level 2 autonomy driving features. Canalys. https://www.canalys.com/newsroom/canalys-level-2-autonomy-vehicles-US-Q2-2019?time=1611158196. Accessed 20 Jan 2021
3. Xie, J., Doshi, R., Tan, J., Jones, C., Ryan, M.: US automotive market Q1 2020. Canalys. https://www.canalys.com/newsroom/Canalys-US-intelligent-vehicle-analysis-Q1-2020?time=1611158459. Accessed 20 Jan 2021
4. Mersky, A.: Near-Term Impacts of Automated Vehicle Technologies. American Council for an Energy-Efficient Economy, Washington DC (2021)
5. NAS: Assessment of Technologies for Improving Light-Duty Vehicle Fuel Economy—2025–2035. National Academies of Sciences, Engineering, and Medicine, Washington, DC (2021)
6. DOE: SMART Mobility: Connected and Automated Vehicles Capstone Report. U.S. Department of Energy (2020)
7. Han, J., Karbowski, D., Kim, N.: Closed-form solutions for a real-time energy-optimal and collision-free speed planner with limited information. Presented at the 2020 American Control Conference (ACC) (2020)
8. Shen, D., Han, J., Jeong, J., Karbowski, D., Rousseau, A.: Receding horizon reference governor for implementable and optimal powertrain-aware eco-driving. IFAC-PapersOnLine **53**(2), 13842–13849 (2020). https://doi.org/10.1016/j.ifacol.2020.12.895
9. ACM: American Center for Mobility | Connected. Automated. Validated. American Center for Mobility. https://www.acmwillowrun.org/. Accessed 30 Sept 2021
10. Hardman, S., Chakraborty, D., Kohn, E.: A quantitative investigation into the impact of partially automated vehicles on vehicle miles travelled in California (2021).https://doi.org/10.7922/G2XK8CVB
11. Wadud, Z., MacKenzie, D., Leiby, P.: Help or hindrance? The travel, energy and carbon impacts of highly automated vehicles. Transp. Res. Part A Policy Pract. **86**, 1–18 (2016). https://doi.org/10.1016/j.tra.2015.12.001
12. Brown, E.G.: Executive Order B-55-18 To Achieve Carbon Neutrality (2018)
13. Ganson, C.: Automated Vehicle Principles for Healthy and Sustainable Communities. California Multi-Agency Workgroup on AV Deployment for Healthy and Sustainable Communities (2018)
14. CARB: Clean Miles Standard. California Air Resources Board. https://ww2.arb.ca.gov/our-work/programs/clean-miles-standard. Accessed 23 Sept 2021
15. Sarmah, D.D.: Autonomous vehicles: zero emissions. Auto Tech. Rev. **4**(6), 1 (2015). https://doi.org/10.1365/s40112-015-0921-x

16. Greenblatt, J.B., Saxena, S.: Autonomous taxis could greatly reduce greenhouse-gas emissions of US light-duty vehicles. Nat. Clim. Change (2015). Advance online publication. https://doi.org/10.1038/nclimate2685
17. VIA: Public Mobility Solutions. Via Transportation. https://ridewithvia.com/. Accessed 23 Sept 2021
18. He, C.R., Ge, J.I., Orosz, G.: Fuel efficient connected cruise control for heavy-duty trucks in real traffic. IEEE Trans. Control Syst. Technol. **28**(6), 2474–2481 (2020). https://doi.org/10.1109/TCST.2019.2925583
19. Ge, J.I., Avedisov, S.S., He, C.R., Qin, W.B., Sadeghpour, M., Orosz, G.: Experimental validation of connected automated vehicle design among human-driven vehicles. Transp. Res. Part C Emerging Technol. **91**, 335–352 (2018). https://doi.org/10.1016/j.trc.2018.04.005
20. He, C.R., et al.: Improving fuel economy of heavy-duty vehicles in daily driving. Presented at the 2020 American Control Conference (ACC) (2020)
21. Assanis, D., Zhao, L., Malikopoulos, A.A.: Characterization of the new class of driving cycles for connected and automated vehicles. Presented at the 2018 21st International Conference on Intelligent Transportation Systems (ITSC) (2018)
22. Ishtiaque Mahbub, A.M., Zhao, L., Assanis, D., Malikopoulos, A.A.: Energy-optimal coordination of connected and automated vehicles at multiple intersections. Presented at the 2019 American Control Conference (ACC) (2019)

Sharing Automated Vehicles: Policies and Ideas to Improve the Sharing Experience to Reduce Congestion and Energy Use in a Post-COVID World

Joshua Auld[1], Denise Baker[2], Danielle Chou[3], Ann Foss[4(✉)], Elizabeth Machek[5], William Riggs[6], and Scott Smith[5]

[1] Argonne National Laboratory, 9700 S. Cass Avenue, Lemont, IL 60439, USA
jauld@anl.gov
[2] Arizona State University, 1151 S. Forest Avenue, Tempe, AZ 85281, USA
denisesbaker@outlook.com
[3] Department of Energy/AAAS Science and Technology Policy Fellowship, 1000 Independence Avenue SW, Washington, DC 20585, USA
[4] City of Arlington, 101 W. Abram Street, Arlington, TX 76010, USA
ann.foss@arlingtontx.gov
[5] Volpe National Transportation Systems Center, 55 Broadway, Cambridge, MA 02142, USA
{elizabeth.machek,scott.smith}@dot.gov
[6] University of San Francisco, 2130 Fulton Street, San Francisco, CA 94117, USA
wriggs@usfca.edu
https://www.linkedin.com/in/danielle-chou

Abstract. Research suggests widespread proliferation of automated vehicles (AV) can potentially greatly increase transportation energy use and congestion [1]. One of the ways to mitigate such increases is to increase sharing in order to provide more environmentally and financially sustainable and cost effective services that match consumer demands for reliability and convenience. This chapter explores how sharing can be encouraged through economic, technological, procedural/legal, and cultural levers in order for AV transportation systems to reduce energy use and congestion.

Keywords: Automated vehicles · AV · Transportation system · Sharing · TNC · Ride-hail · Car-share · Sustainability · Energy · Congestion

1 Introduction

Since the mid-twentieth century, American transportation systems have been centered around private vehicle ownership. Multiple factors contributed to this, ranging from urban planning trends and the construction of the highway systems to the availability and affordability of vehicles to middle income families. These structural and cultural inclinations have been further reinforced by the COVID-19 pandemic, during which people have been urged to social distance and isolate. Private vehicle ownership increased

G. Meyer and S. Beiker (Eds.): ARTSymposium 2021, LNMOB, pp. 71–82, 2023.
https://doi.org/10.1007/978-3-031-11112-9_7

during the pandemic [2], and transit ridership tumbled to its lowest levels in 20 years [3].

These trends may have a strong influence on the burgeoning automated vehicle (AV) industry, for which operational and business models are still being developed. New innovations in automated driving offer the potential to enhance accessibility and mobility, thus fueling economic activity and development. Nevertheless, the pivot towards automated driving may result in negative externalities and exacerbate existing transportation challenges, including air pollution, congestion, carbon emissions, and long commutes. The balance of these outcomes will likely be heavily influenced by operational models: in the best case scenario, publicly shared AVs result in less deadheading, so fewer vehicles are needed to fulfill travel demand, and the system would therefore reduce energy use and congestion. In the worst case scenario, privately shared AVs are dispatched to meet only the travel demands of their own households and thereby incur many deadheading miles to do so. The system thus increases energy use and congestion.

These two scenarios also utilize different ownership and business models: for publicly shared AVs, the vehicles may be privately owned by individuals, privately owned by companies, or publicly owned. This reflects the transportation network company (TNC), car-share, and public transit business models, respectively. In contrast, privately shared AVs use the private ownership model that dominates the U.S. transportation system today.

There are a range of choices that community planners, engineers, and policy makers can make to steer trends in AV deployment towards more convenient, reliable, and sustainable mobility outcomes. In this chapter, we explore operational and business models and summarize an ARTS21 workshop discussing strategies to mitigate the energy and congestion impacts of single occupant AVs and to facilitate shared mobility. We first discuss the potential of sharing journeys and/or vehicle assets to mitigate potential increased energy use and congestion. We then explore public attitudes towards AVs and sharing, with a particular focus on the US market. Finally, we dive into more specific economic, technological, procedural/legal, and cultural levers that can help encourage sharing.

1.1 Motivation for Shared AVs: Energy Use and Congestion

Rapid advances in sensing, onboard computing, machine vision and machine learning have brought automated vehicle technologies to the point of near feasible deployment. The transition from human-driven vehicles to automated vehicles (AV) has been posited to have many overall advantages from a societal perspective ranging from reduced traffic accidents and other safety incidents, increased roadway capacity and throughput, increased productive time use during travel and reduction in driver burden, to reduced emissions and energy use due to the above traffic and congestion improvements as well as more efficient driving cycles, among many other potential impacts [4]. However, these benefits come at potential costs, including increased traffic and congestion as driving burden is reduced, instability in traffic streams due to AV controls, increased energy use due to sensing and computing requirements, as well as possible higher speeds or increased congestion, increased urban sprawl, and so on [4, 5]. The balance between positive and negative impacts largely hinges on the influence of four factors: vehicle design and operation, vehicle ownership modalities, fleet operations, and traveler behavior. The interplay between these factors is quite complex, leading to wide variation in hypothesized outcomes for automated vehicle impacts [1, 5, 6].

Critical questions to address in terms of understanding potential impacts of AV include:

- Are the vehicles privately owned or operated in shared fleets? [7]
- Are the vehicles operated cooperatively in traffic streams or as individuals? [8, 9]
- Does riding in an automated vehicle encourage extra travel (through reduced burden, alternative time use, etc.)? [10–12]
- Are people willing to ride in shared vehicles operated autonomously?

These are open research questions that have been extensively studied in recent years. A 2016 scoping study funded by the US Department of Energy [1] noted the wide potential uncertainties in AV impacts leading to anywhere from a 200% increase to a 60% decrease in fuel usage. This motivated further research through the DOE SMART Mobility research program to use advanced transportation system and vehicle energy simulation tools to further explore the potential bounds of AV impact in more detail. A key unknown addressed in this research was the ownership model that may occur with AV deployment. That is, will vehicles continue to be largely privately owned as is currently the case today, or will the shared use model of mobility (i.e., shared fleets) dominate?

In order to explore the range of outcomes, a variety of shared and automated scenarios for future mobility deployment were developed and simulated in the POLARIS agent-based transportation simulator [12] for the Chicago metropolitan region. The scenarios explored included a high sharing/low automation case (AV limited to partial automation), a high sharing/high automation case (full AV deployed as TNC fleet vehicles), and a low-sharing/high automation case (AV largely privately owned). In both the shared and non-shared cases, AV penetration ranged from 18% to 52%. In both high sharing scenarios, each additional fleet vehicle replaced five private vehicles (for non-AV) and ten private vehicles (for AV fleet vehicles).

The comparison of results across the shared and automated axes demonstrated several key findings. First, it was observed that shared-use led to higher system efficiency: a constant level of mobility (measured in productive miles of travel) was delivered at lower system congestion levels and a 23% decrease in energy use. Meanwhile, privately owned AVs resulted in a slight increase in productive travel, but at the expense of an 18% reduction in speed and 22% increase in energy use. This difference was largely driven by the much greater efficiency of vehicle re-positioning for fleet automated vehicles versus privately owned AVs (which re-position only to accommodate other household members). In fact, in the high penetration private AV case, 1 in every 7 vehicles was operating empty at any given time due to private AVs re-positioning for other household members, versus 1 in 25 for the shared AV fleet. To be clear, these scenarios make many assumptions regarding vehicle adoption, fleet operations and traveler behavior – including open questions around changes in travel time, ride pooling in automated vehicles, long-term household adaptation (i.e., relocation, vehicle disposal, etc.). However, the results indicate that when automated vehicles operate in an efficient shared fleet, they can reduce the amount of deadheading and provide service levels sufficient to allow for private vehicle ownership reductions. Therefore, such vehicle sharing provides a potential solution to mitigate energy impacts of AV deployment.

1.2 Public Attitudes About Sharing

Considering all the sustainability and efficiency benefits of sharing, understanding of public attitudes and adoption of shared modes is critical. In general, consumers have not adopted shared peer-to-peer platforms at a quick pace. Hence, there is ample opportunity to expand this market and shape new attitudes to more readily accept shared rides. Transit vehicles (including bus, paratransit, and rail) and ride-hailing services (such as Uber Pool and Lyft Line) provide the opportunity to share rides between unfamiliar passengers. There are also other forms of sharing such as carsharing and micro-mobility services in which people share vehicles rather than rides.

Based on the National Household Travel Survey results from 2017, transit vehicles including all paratransit, bus, and rail modes account for 4.4% of person miles traveled and 4.7% of person trips. Commuting trips account for half of the transit trips [13]. The COVID Future survey suggests there has been about a 40% decline in transit commute trips post-pandemic, relative to pre-pandemic baselines. Of this decline, about half can be attributed to changes in commuting frequency, 40% from a net shift among transit commuters toward the private car, and the remaining 10% from shifts to other modes [14].

Ride-hailing services such as Uber and Lyft provide on-demand mobility services to people using cellphone ride-hailing apps wherever the service is available. From 2009–2017, the for-hire vehicle mode share doubled based on National Household Travel Survey data. As of 2017, for-hire vehicles still only account for 0.5% of all trips, and the percent of all Americans who use ride-hailing in any given month is nearly 10% [15]. This trend of growth has been greater in mid-sized and large cities and among younger individuals and wealthier households [15]. A national survey found that only 7% of ride-hailing users combine ride-hailing trips with public transit on at least a weekly basis, while 35% do so at least occasionally [16].

While ride-hailing services have not yet become a popular mode of transportation among all segments of the population, they demonstrate a potential operating model for future automated transportation systems in which mobility needs are meant on demand and with a possible expectation of sharing. This model is justified by the market economics of most automated vehicle companies, which would require shared rides to increase the number of people served and revenues produced per people-mile traveled [17]. Therefore, investigating current perceptions around sharing and the rate of sharing on existing ride-hailing platforms across different population groups can help decision makers better plan for the future of shared automated mobility.

The T4 survey which was conducted in four southern metro areas in the US during 2019 (pre-pandemic) has shown that while 16% of the weighted respondents (n = 3358) are using Uber and Lyft in private mode at least monthly, only 7% are using UberPool and Lyft Line (shared mode). Investigating the perceptions around sharing, 46% of respondents stated that traveling with a driver they don't know makes them feel uncomfortable while 61% stated that traveling with unfamiliar passengers makes them feel uncomfortable. Additionally, only 26% believe that the lower cost of shared ride-hailing is worth the additional time picking up and dropping off other passengers. This finding implies that more work should be done on effective pricing algorithms for shared rides to be economically persuasive compared to private rides [18].

The same T4 survey asked respondents to report their last ride-hailing trips. Out of 1219 reported actual ride-hailing trips, only 12% chose to share (in Austin and Atlanta where the shared option was available). Concerning the socioeconomic attributes of the traveler, low income riders shared at a rate twice that of high income riders; women shared 1.5 times more than men; and, frequent users shared 1.4 times more than infrequent users [18]. In another study, Sarriera et al. found that younger, unmarried, non-car-owning individuals were more likely to share their ride-hailing trip; income and gender did not have a significant effect on ridesharing opt-in; and the majority of ridesharing trips were for leisure, rather than commuting or airport access [19].

The T4 survey also asked the respondents' about their potential willingness to share automated ride-hailing trips in the future. While 46% stated that they would use AVs alone or with family, friends, and coworkers, only 20% stated that they would share their AV rides with unfamiliar passengers. Stated willingness to share rides in AV decreases among older individuals, females, and very high-income respondents. The T4 survey also proved that experience matters. Existing users of shared ride-hailing services were twice as willing to share an AV ride with strangers. Another important factor in willingness to share besides socioeconomic characteristics and real experiences is attitudes. People with environmentally-friendly and tech-savvy attitudes were more willing to share an AV ride in the future with unfamiliar people. Likewise, the COVID-19 pandemic has certainly imposed certain changes in people's mindsets toward sharing [18].

In summary, existing policy, security, and pricing systems have not been very successful at shifting people towards sharing. While we should continue to enhance our policy, security, and pricing incentives, it is critical that we work on creating positive attitudes and perceptions about sharing so that individuals decide to share for the benefit of themselves and the society that they live in. We have to make shared options attractive in terms of price; comfortable in terms of security; safe in terms of mitigating disease transmission; efficient in terms of accessibility, travel time, and connectivity to transit; and informative with rich communication of the benefits of sharing. We should not forget that experience matters, and pilot projects to provide free shared rides could prove the effectiveness of shared rides to users.

2 Levers Impacting the Adoption of Shared AVs

As a result of ARTS21 workshop activities, several levers were identified which could impact the adoption of shared AVs. These levers can be roughly categorized into 4 groups: economic, procedural/legal, technological, and cultural levers. While these categories are not meant to be exhaustive nor comprehensive, they provide a framework for thinking about how those in policy planning and engineering can facilitate greater adoption of shared vehicles.

2.1 Economic Levers

First and foremost, to make shared AVs financially sustainable, there must be a business case for them. It is critical to understand who the customer is, yet in this nascent market, there appears to be some hesitation to question existing business models. OEMs are experienced at selling to customers, so in the absence of other good examples, that is the business model they will continue to pursue – and this is not particularly conducive to sharing.

Alternative business models other than private vehicle ownership can be challenging. Taxis and some car-sharing models can be very capital intensive: they require vehicle procurement as well as securing parking somewhere. In this regard, they can struggle to compete against transportation network companies (TNCs) such as Uber and Lyft, which reduce those capital costs by leveraging existing vehicle assets. For all of these business models, population and activity density are critical to improving vehicle utilization.

To be financially viable, shared AVs must offer benefits over personally owned vehicles. These incentives can be cost (cheaper), speed (reduced travel time), and/or user experience. Incentives are particularly important for low-income families, where cost is a major consideration. In 2020, the average US household spent approximately 16% of its income on transportation [20]. How will we maintain equitable access to AVs? It is possible more affluent families may decide to either use a shared AV as their secondary vehicle or rent out their secondary AV to a TNC service. AVs may also start as an elite luxury service; this type of business case would also make it unlikely that passengers would be willing to share such an amenity with strangers. Nudging travel behaviors can also be done by disincentivizing single occupant vehicle trips with tools such as congestion pricing or expensive parking.

Financial risk aversion can stem from operational uncertainties, as well as low confidence in consumer adoption. Operational uncertainties include how to enforce unsupervised fare collection from every passenger (should the vehicle stop if there is a free rider?) and how to manage such an operation at scale. This is where small-scale demonstrations can be critical to acclimating prospective riders and operators; this topic is discussed at greater length in the Cultural Levers section.

2.2 Procedural/Legal Levers – Trust

To encourage sharing, trust within the community is critical. Transparent and consistent procedural and legal levers must work to advance that trust. There are two forms of trust inherent in shared AV services: trust in technology and trust in fellow riders.

For trust in technology, a major adoption barrier for sharing AVs is trust across the public and private sectors and between consumers and technology providers. Education and outreach are necessary levers to expose and acclimate the public to these technologies. Hence, pilot deployments play a critical role in defining the public perception of AVs. Again, these deployments should be carefully tailored to use cases where they would gain the most public traction and can start small to build confidence over time. For some cities, this may be providing a feeder service for transit. For other cities, shared AVs may be the primary transit option. At the same, as the public acclimates to these technologies, successful deployments can bolster the confidence of prospective operators when they see how AVs can be used to provide reliable, useful service and how to incrementally overcome operational issues.

Additionally, government regulatory review of AV technologies may also help increase consumer confidence. As the technology evolves and motor vehicle safety standards specific to automation are issued, consumer confidence may also increase.

To facilitate trust in fellow riders, rider identity authentication was flagged as a critical technology gap for shared AVs. Its resolution is required for the public to feel the system is safe and fair. Without a human driver in a supervisory role, lack of trust in fellow passengers could be a significant obstacle to journey sharing in small automated vehicles. With transit, riders can change train cars or take another bus when they feel unsafe. That will probably not be the case for shared AVs whose riders are matched to align journeys rather than follow pre-defined routes. It can also ensure that fares are correctly collected. Likewise, rider identity authentication can be useful to prospective fleet operators who may be wary of potential liability should an incident or assault happen within their unsupervised vehicles. It could help companies more clearly understand their legal liability in unsupervised travel settings and could help the public hold the appropriate entities accountable. Rider identification also enables rider accountability. It would be possible to track undesirable behaviors that affect other riders, ranging from being late, to using someone else's fare payment card, to assaulting other riders. Measures could range from warnings to being banned to arrest and prosecution.

Several procedural actions could be taken to make the shared journey more secure, pre- during- and post-trip. First is to make sure riders are identified and use the correct vehicle. This can be done via:

- Pre-registration for the service;
- Payment via a means that also verifies identify (e.g., contactless payment card, which could be a credit card or card that is dedicated to transportation payments, such as a transit smart card);
- Tapping in via this card will make it less likely the rider boards the wrong vehicle (or a fake TNC vehicle).

Pre-trip, there may be selectivity on journey matching by type of rider (following the example of women-only minicabs, e.g., Lady Cabs). However, this may have a cost in journey-matching efficiency.

2.3 Procedural/Legal Levers – Land Use

Local land use and zoning regulations also have an impact on the efficiency of shared vehicles. Network design (hub-to-hub versus door-to-door) may make it easier for a larger number of people to share journeys, while keeping circuitous travel to a minimum. Land use policies that encourage greater density in housing and activity opportunities make transportation more efficient in general and would help business cases be more viable as well.

Increased density can increase use of more active modes (walking, bicycling, micromobility). They can also address negative externalities that regularly surface in dystopian visions of autonomous futures by helping to curb urban sprawl, congestion, emissions, and long commuting patterns – all of which have major social, environmental, and economic impacts [21–23]. However, since the 1950s, land use policies have trended towards decreased density (including increased road widths, minimum home lot sizes, and single use zoning districts), which support private vehicle travel modes over shared, public, and active travel modes. It is unclear if these trends will be reversed.

2.4 Technological Levers – Vehicle and Infrastructure Design

There are several vehicle design factors which could impact the acceptability of sharing journeys, particularly for groups that include strangers. Previous research has identified several design factors which contribute to the acceptability of shared interior spaces. During the trip, physical design of the vehicle may enhance security. Similar to automated guideway vehicles today, vehicles can have security cameras and call buttons. With larger vehicles (e.g., the size of a transit bus), security may be enhanced by the presence of other riders. Other vehicle design considerations include the aspects outlined in the sections that follow.

2.4.1 Seating Configuration and Ease of Entry and Exit

Vehicles which are designed to accommodate groups of passengers boarding and alighting at different stops would ideally allow for an individual to smoothly enter, navigate to a seat, and then later leave that seat, without close physical proximity to another passenger or being impeded by another passenger or another passenger's belongings. Bench (longitudinal) seating is generally faster to load and unload than side-by-side (transverse) seating. There are anecdotal reports that passengers generally prefer to face forward, but at least one observational study has found no clear preference [24]. Seating configuration tradeoffs have been well-studied [25] for traditional transit vehicles (bus, subway, etc.) although the primary decision there is often to balance the space available between seated passengers and standees. Seating configuration can also enhance personal space and privacy. For example, a conceptual vehicle could place seats back-to-back so that passengers are looking out of a window instead of at the face of a stranger, but these needs would need to be assessed against safety and crashworthiness.

2.4.2 Cargo Space

Providing space on board the vehicle for passenger belongings can help reduce the friction caused by competition for limited space between passengers.

2.4.3 Vehicle Size

While smaller vehicles offer flexibility for serving dispersed destinations, the smaller interior space also creates an intimacy of experience which may increase passenger discomfort for groups including strangers. Further study is likely required to determine the most appropriate vehicle size, but it seems plausible that slightly larger vehicles offering more personal space would increase passenger comfort.

With small vehicles, separate compartments (or at least dual armrests to mark personal space) could be provided. Compartments could be designed to hold one rider, plus baggage (or, possibly space for a child in a child seat). Such a compartment could also be roomy enough for a wheelchair [26].

2.4.4 Child Seats

If a conventional passenger vehicle is used, providing a booster seat or other child seat expands usability for those traveling with small children.

There also are infrastructure-based levers, including the use of higher-occupancy lanes, preferential parking, or signal priority for vehicles making shared journeys to make shared vehicles a more efficient option. These additional civil infrastructure strategies offer opportunities to think about positive sum roadway design, which is a factor that has long been discussed at TRB conferences and other planning events. For example, everything from curbs to sidewalks to the streets could be thought of in order to impact ingress and egress at the curb [27]. Ultimately, streets could be reconfigured with less emphasis on automobiles and more space for cycling, walking, and automated transit [28, 29].

2.5 Cultural Levers

There are also potential cultural barriers to the use of AVs and shared AVs in particular; other levers can potentially influence these barriers. Three key categories of cultural levers include:

- selecting the right use case for the situation;
- using appropriate framing for education and outreach;
- demonstrating how shared AVs can successfully meet rider needs.

Selecting the right use case includes both riders and trips – identifying who is most likely to benefit from shared AVs and when riders can benefit from shared AVs. Shared trips can be less expensive, so, as might be expected, lower income riders might be more likely to appreciate shared options. Additionally, riders may be more open to shared rides for certain types of trips, such as travel to/from an airport where riders are already in the mindset of sharing travel space [30]. When shared AV trips are safer, lower cost, and accessible to all users – in short, more efficient – they are more likely to be selected. In addition to certain trips being conducive to shared rides, certain areas might also be more conducive. A shared ride in a dense and lively area, where riders are frequently boarding or being dropped off, might be more appealing than a long trip with just one other unknown rider.

Culturally appropriate framing is critical for successful implementation of shared AVs. If the target audience and use case is established, then education and outreach campaigns can be tailored to best reach potential riders. Suitable frames may include economic development – how a shared AV service is good for business, tourism, etc.; safety – how AVs can reduce roadway crashes and fatalities; or environmental benefits – how shared AVs can reduce emissions, congestion, or need for parking lots. Once the framing messages are identified, ample educational materials and outreach opportunities should be provided to allow potential riders to learn about the service, the technology, and the benefits of shared AVs.

A key component of education and outreach is allowing the public to experience shared AV service through pilot and demonstration programs. Pilots should be designed to address the challenges of mode shift, getting potential riders out of their normal mode and into the shared AV to try the service once through free rides or other incentives. For example, during a bus rapid transit (BRT) adoption and demand response study in Australia, the BRT services initially operated at a financial loss, but the new service became economically viable within six months through outreach, education, and word of mouth. The service now provides benefits without requiring heavy subsidies.

Incentives can also be used to gamify the travel experience, as providing riders with points or credits can help shift travel behavior. If shared rides are the end result, it may be easier to start with shared rides from the beginning, so as to avoid a shift in expectations. A survey conducted in 2017 found that riders who are accustomed to solo ride-hailing services are less likely to select a shared ride [31], similar to what was found with Uber Pool and Lyft Line services. Thus, establishing a norm of shared AV rides in pilot demonstrations may be critical during this formative period for the budding AV industry.

3 Conclusions

Research suggests the emerging AV industry has the potential to improve mobility and accessibility. At the same time, it has the potential to increase congestion and energy consumption. One of the ways to mitigate increases in energy use is to promote sharing of either vehicle assets or journeys, but widespread sharing faces many barriers in the US given the strong proclivity for private vehicle ownership, which has been further reinforced by the COVID-19 pandemic.

Efforts to promote sharing should be targeted towards specific use cases where the technology, design, and economics are appropriate for the audience in question. Just like AVs, sharing may not work everywhere. In particular, population density is a major design driver – and at this point, it is unclear whether pandemic-induced trends in favor of teleworking and larger homes further from cities are transient or will become permanent. Exurban and lower-density growth does not preclude the use of shared AVs, but it has significant implications for energy consumption and overall vehicle miles traveled (VMT).

The current COVID-19 pandemic-induced housing and labor market upheaval is potentially upending assumptions about land use, and it is unclear what trends will persist long term. Yet breaking down barriers to sharing is critical regardless of this uncertainty. There are technological, procedural/legal, economic, and cultural levers that can help alleviate some of these barriers, including: developing shared vehicle designs that are easy to use, clean, and provide some measure of privacy; increasing trust in technology through education and exposure to truly beneficial use cases; increasing trust in fellow riders by developing rider authentication processes; building infrastructure that favors and prioritizes shared modes to make them faster and more convenient than other alternatives; and formulating pricing schemes that incentivize sharing. Given the AV industry and associated operational models are still nascent, further research in this area can influence what form our AV future takes to create a more reliable, convenient, environmentally sustainable, and equitable transportation system.

References

1. Stephens, T.S., Gonder, J., Chen, Y., Lin, Z., Liu, C., Gohlke, D.: Estimated bounds and important factors for fuel use and consumer costs of connected and automated vehicles. Technical report NREL/ TP-5400-67216; National Renewable Energy Laboratory, Golden, CO (2016)
2. https://data.bts.gov/Research-and-Statistics/Auto-Sales/7n6a-n5tz Accessed 10/28/21
3. https://data.bts.gov/Research-and-Statistics/Transit-Ridership-Urban-Rail/rw9i-mdin. Accessed 28 Oct 2021
4. Taiebat, M.; Brown, A.L., Safford, H.R., Qu, S., Xu, M.: A review on energy, environmental, and sustainability implications of connected and automated vehicles. Environ. Sci. Technol. **52**, 11449–11465 (2018)
5. Wadud, Z., MacKenzie, D., Leiby, P.: Help or hindrance? The travel, energy and carbon impacts of highly automated vehicles. Transp. Res. Part Policy Pract. **86**, 1–18 (2016)
6. Brown, A., Gonder, J., Repac, B.: An analysis of possible energy impacts of automated vehicle. In: Meyer, G., Beiker, S. (eds.) Road Vehicle Automation. LNM, pp. 137–153. Springer, Cham (2014). https://doi.org/10.1007/978-3-319-05990-7_13
7. Fagnant, D.J., Kockelman, K.M.: The travel and environmental implications of shared autonomous vehicles, using agent-based model scenarios. Transp. Res. Part C Emerg. Technol. **40**, 1–13 (2014)
8. Mahdinia, I., Arvin, R., Khattak, A.J., Ghiasi, A.: Safety, energy, and emissions impacts of adaptive cruise control and cooperative adaptive cruise control. Transp. Res. Rec. **2674**(6), 253–267 (2020). https://doi.org/10.1177/0361198120918572
9. Liu H., Lu, X.Y., Shladover, S.E.: Mobility and energy consumption impacts of cooperative adaptive cruise control vehicle strings on freeway corridors. Transp. Res. Rec. (2020)
10. Harb, M., Xiao, Y., Circella, G., Mokhtarian, P.L., Walker, J.L.: Projecting travelers into a world of self-driving vehicles: estimating travel behavior implications via a naturalistic experiment. Transportation **45**(6), 1671–1685 (2018). https://doi.org/10.1007/s11116-018-9937-9
11. Pudane, B., Molin, E.J.E., Arentze, T.A., Maknoon, Y., Chorus, C.G.: A time-use model for the automated vehicle era. Transp. Res. Part C Emerging Technol. **93**, 102–114 (2018)
12. Auld, J.A., et al.: Exploring the mobility and energy implications of shared versus private autonomous vehicles*. IEEE Intell. Transp. Syst. Conf. (ITSC) **2019**, 1691–1696 (2019). https://doi.org/10.1109/ITSC.2019.8917125
13. Federal Highway Administration; National Household Travel Survey, version 1.2, Federal Highway Administration (2017)
14. Salon, D., et al.: The potential stickiness of pandemic-induced behavior changes in the United States. Proc. Nat. Acad. Sci. **118**(27), e2106499118 (2021). https://doi.org/10.1073/pnas.2106499118
15. Conway, M.W., et al.: Trends in taxi use and the advent of Ridehailing, 1995–2017: evidence from the US national household travel survey (2018)
16. https://www.masabi.com/. Accessed 28 Oct 2021
17. Riggs, W., Beiker, S.A.: Business models for shared and autonomous mobility. In: Meyer, G., Beiker, S. (eds.) AVS 2019. LNM, pp. 33–48. Springer, Cham (2020). https://doi.org/10.1007/978-3-030-52840-9_4

18. https://tomnet-utc.engineering.asu.edu/t4-survey/. Accessed 28 Oct 2021
19. Sarriera, J.M., Álvarez, G.E., Blynn, K., Alesbury, A., Scully, T., Zhao, J.: To share or not to share: investigating the social aspects of dynamic ridesharing (2017)
20. U.S. Department of Transportation, Bureau of Transportation Statistics, Transportation Economic Trends. www.bts.gov/product/transportation-economic-trends. Accessed 28 Oct 2021
21. Riggs, W., Larco, N., Tierney, G., Ruhl, M., Karlin-Resnick, J., Rodier, C.: Autonomous vehicles and the built environment: exploring the impacts on different urban contexts. In: Meyer, G., Beiker, S. (eds.) AUVSI 2018. LNM, pp. 221–232. Springer, Cham (2019). https://doi.org/10.1007/978-3-319-94896-6_19
22. van Arem, B., Ackerman, A., Chang, T., Riggs, W., Wegscheider, A., Smith, S., Rupprecht, S.: Building Automation into Urban and Metropolitan Mobility Planning. In: Meyer, G., Beiker, S. (eds.) AVS 2018. LNM, pp. 123–136. Springer, Cham (2019). https://doi.org/10.1007/978-3-030-22933-7_13
23. Riggs, W.: Disruptive Transport: Driverless Cars, Transport Innovation and the Sustainable City of Tomorrow. Routledge, London (2019)
24. Berkovich, A., Lu, A., Levine, B., Reddy, A.V.: Observed customer seating and standing behavior and seat preferences on board subway cars in New York City. Transp. Res. Rec. **2353**(1), 33–46 (2013). https://doi.org/10.3141/2353-04
25. O'Malley, T., Vaishnav, M.: Is this seat taken?: Multifaceted research study to inform the Chicago, Illinois, transit authority's future rail car seating design. Transp. Res. Rec. **2419**(1), 33–39 (2014). https://doi.org/10.3141/2419-04
26. Merat, N., Madigan, R., Nordoff, S.: Human factors, user requirements, and user acceptance of ride-sharing in automated vehicles. Discussion Paper: International Transport Forum, Paris (2017)
27. Riggs, W., Ruhl, M., Rodier, C., Baumgardner, W.: Designing streets for autonomous vehicles. In: Road Vehicle Automation, vol. 6, pp. 111–122 (2019)
28. Riggs, W.: Technology, civic engagement and street science: hacking the future of participatory street design in the era of self-driving cars. In: Proceedings of the 19th Annual International Conference on Digital Government Research: Governance in the Data Age, Delft, Netherlands, pp. 4:1–4:6 (2018). https://doi.org/10.1145/3209281.3209383
29. Schlossberg, M., Riggs, W.W., Millard-Ball, A., Shay, E.: Rethinking the street in an era of driverless cars. UrbanismNext (2018). https://urbanismnext.uoregon.edu/files/2018/01/Rethinking_Streets_AVs_012618-27hcyr6.pdf
30. Hou, Y., Garikapati, V., Weigl, D., Henao, A., Moniot, M., Sperling, J.: Factors influencing willingness to pool in Ride-Hailing trips. Transp. Res. Rec. J. Transp. Res. Board (2020). https://doi.org/10.1177/0361198120915886
31. Lavieri, P.S., Bhat, C.R.: Mobility-as-a-service in car-dominated cities modeling individuals' willingness-to-share trips with strangers in an autonomous vehicle future. In: Innovation in Travel Modeling Conference, Atlanta, GA (2018). http://onlinepubs.trb.org/onlinepubs/Conferences/2018/ITM/PLavieri.pdf

Part III: Users and Human Factors

Human-Centric Intelligent Driving: Collaborating with the Driver to Improve Safety

Avinash Balachandran[1(✉)], Tiffany L. Chen[1], Jonathan Y. M. Goh[1], Stephen McGill[2], Guy Rosman[2], Simon Stent[2], and John J. Leonard[2]

[1] Toyota Research Institute, 4440 El Camino Real, Los Altos, CA 94022, USA
{avinash,tiffany.chen,jonathan.goh}@tri.global
[2] Toyota Research Institute, Building 100, Suite 1-201, Cambridge, MA 02139, USA
{stephen.mcgill,guy.rosman,simon.stent,john.leonard}@tri.global

Abstract. Despite the benefits of autonomous vehicles, their many challenges have made their wide scale deployment and adoption slower than hoped for. In order to help spread the potential benefits of autonomy sooner, as well as to cater to people who will continue to prefer to drive themselves while improving safety, there is a need for intelligent interaction and collaboration between increasingly automated vehicles and humans. At Toyota Research Institute, we call this Human-Centric Intelligent Driving (HCID). HCID has many technical challenges, some of which are shared with fully autonomous driving. Due to the collaborative nature between humans and machines in HCID, some of these challenges are particularly important and potentially different from fully autonomous driving. This chapter focuses on Toyota Research Institute's (TRI) approach to addressing some of these core challenges.

Keywords: Autonomy · Human-centric · Driver monitoring · Driving risk modeling · Behavioral forecasting · Intelligent transportation · Model predictive control · Vehicle dynamics

1 Introduction

Autonomous vehicles promise a litany of potential benefits from increased safety to reduced congestion and even environmental benefits [1, 2]. The race for autonomous vehicles also highlights the numerous challenges to its development. Not only are there core technical challenges such as getting more accurate predictions or ensuring rigorous testing/validation, non-technical challenges also exist, such as market introduction and regulatory frameworks [3, 4]. As academia, industry, and the public sector work to address these challenges, wide scale deployment and adoption of fully autonomous vehicles has been slower than hoped for [5]. This has created a growing need for intelligent interaction and collaboration between increasingly, but not fully, automated vehicles and humans. Such a collaborative system not only promises a potentially quicker path towards reaping some of the benefits of autonomous vehicles like increased safety, it also may enable faster adoption of truly autonomous vehicles by increasing human acceptance and trust [6].

G. Meyer and S. Beiker (Eds.): ARTSymposium 2021, LNMOB, pp. 85–109, 2023.
https://doi.org/10.1007/978-3-031-11112-9_8

One of Toyota's approaches to building this kind of collaborative system between increasingly automated vehicles and humans is Human-Centric Intelligent Driving (HCID). The HCID concept seeks to use the Artificial Intelligence (AI) being developed for fully autonomous vehicles to support, augment and amplify human driving capability. Figure 1 shows how the HCID concept compares to traditional forms of ADAS. Unlike traditional ADAS, such as Automatic Emergency Braking (AEB) or Driver Monitoring System (DMS), HCID aims to learn a joint representation of the situation across the vehicle, environment and driver to inform its policies. A policy in this context describes the actions taken by the system in response to the representation of the situation. It ingests the full spectrum of inputs from the vehicle (e.g. speed, engine torque etc.), environment (e.g. lane boundaries, cyclists etc.) and human (e.g. steering, throttle, gaze etc.) to build rich joint situational representations. It then generates policies and interactions based on these representations to amplify the human in the driving task, with a goal of improving safety outcomes. This contrasts with traditional ADAS systems where a small set of inputs generates narrower policies and interventions for a specific use case. For example, many AEB systems traditionally mainly consider the driver's brake input and the presence of an obstacle in front of the vehicle to produce a warning and/or hard brake event as necessary to prevent a collision or reduce impact force. The system does not have a deep understanding of the environment (e.g. occupancy of adjacent lanes) or the driver's intent. Therefore, if a skilled and engaged driver executes an aggressive emergency lane change into an empty adjacent lane to avoid an obstacle, the AEB system may be unaware of this and may even have a detrimental impact. For instance, the system could produce a hard brake event or distracting warning during the execution of the emergency lane change, resulting in a loss of concentration or control for the driver. An HCID system, with its rich joint representation, potentially avoids issues like this and gives the best support to the driver, made with the best available knowledge of the situation.

Figure 1 also highlights another difference between HCID and ADAS. In a typical ADAS constellation, each additional ADAS feature is added in parallel to the others resulting in a complex system of features. The interactions between these individual ADAS features can result in unexpected outcomes to the driver. For example, in the emergency lane change situation described above, the Electronic Stability Control (ESC) and/or Anti-Lock Braking System (ABS) system could also engage together with AEB resulting in even more unpredictable dynamics and a greater disconnect between the driver's intent and the vehicle's action. By using its joint representation to build policies, the HCID system can potentially avoid these issues by ensuring consistent policy feedback to the driver based on a holistic understanding of the situation.

Since the HCID system is generating policies based on holistic and rich situational representations, it leverages many of the components of fully automated driving including motion planning/control, perception, prediction, localization and human-machine interaction (HMI). To understand what capabilities are potentially needed for an HCID system, it is instructive to look at the breakdown in driver-related critical reasons for driving accidents: 41% are due to recognition error, 33% are due to decision error, 11% are due to performance error, and 15% are due to driving incapability [7]. An HCID system needs technologies that address and reduce these errors by using information

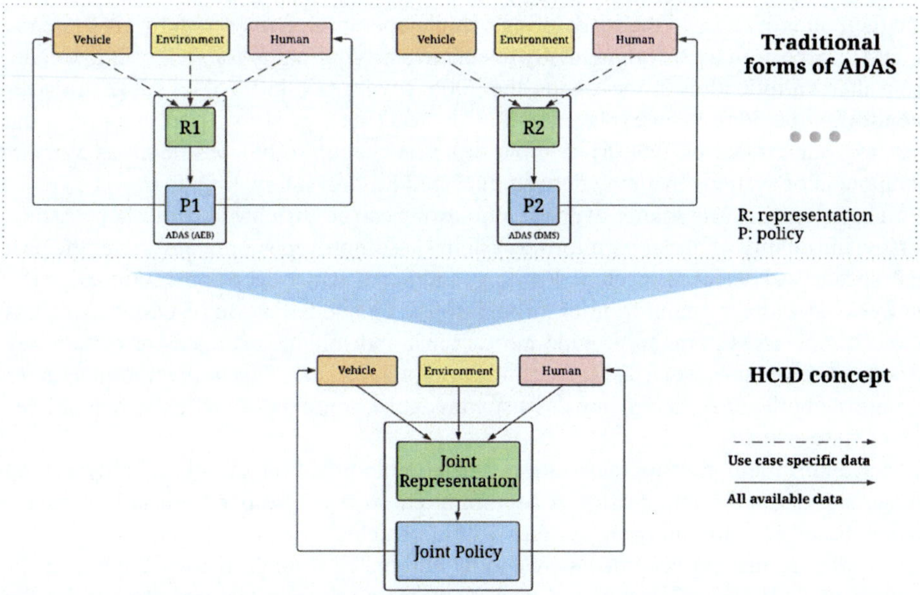

Fig. 1. How the Human-Centric Intelligent Driving (HCID) concept compares to traditional forms of ADAS.

fully available from the environment, vehicle, and the driver. Therefore, the system will ideally have superior situational awareness to amplify scene recognition/understanding and excellent driving skills to augment the driver as necessary to take the best decisions and control actions. Beyond that, the system would also ideally have a good understanding of the driver's intent and state as well as a framework to effectively communicate its policies to the driver to ensure the best outcomes. Therefore, the ability to build superior situational awareness, having expert driving skills, and understanding/interacting with the driver are pillars of the HCID system.

This chapter focuses specifically on some of the core technical challenges to HCID and the several strands of research at the Toyota Research Institute (TRI) addressing them. The research presented can help inform not only HCID but also fully autonomous driving systems and driver amplification systems like Toyota Guardian™ [8, 9]. Section 2 investigates the challenges of driver monitoring and understanding the driver state which helps ensure an HCID system is aligned with the driver's intent. Section 3 looks at building expert skills into the HCID system so that the vehicle is able to amplify and support regular drivers when necessary. Section 4 covers scene understanding as it pertains to risk estimation, modeling, and prediction of human behavior on the road. Section 5 discusses next steps related to building the paradigm for human-machine interactions for the HCID system. Finally, Sect. 6 concludes with a summary of the work discussed.

2 Driver Monitoring

Driver monitoring refers to the use of in-cabin information, often extracted using computer vision techniques from a driver-facing camera, to measure the state of the driver and their attentiveness to the road during vehicle operation. Driver monitoring systems, or DMS, have been used commercially to enhance driving safety for well over a decade, from their introduction in the US on the 2006 Toyota GS 450 h. The value of driver monitoring has long been explored and evaluated (see e.g. [10–13]). It is only in the last few years however, that driver monitoring has begun to be considered as a major component of systems that may deliver enhanced driving safety.

The increasing importance of driver monitoring can be attributed to three key reasons. Firstly, the quality of driver monitoring systems has improved significantly over the past half decade. Advances in machine learning, and in particular deep convolutional neural networks [14], have begun to filter through to the embedded world of DMSs. This has allowed modern systems to provide more robust and reliable estimates of certain key aspects of driver state, such as 3D face and gaze tracking [15]. This is an important factor in improving the error rate in any downstream decision-making function which utilizes driver state estimates.

Secondly, with the miniaturization and commoditization of high-quality digital image sensors and computers, it is becoming easier and cheaper to install computer vision–based driver monitoring systems within vehicles.

Thirdly, as driving becomes increasingly automated through the widespread introduction of SAE Level 2 and early Level 3 automated driving features, the role of the driver as a driving supervisor has begun to receive more attention. Humans can quickly lose vigilance for monotonous tasks where the perceived pay-offs are low [16]. This effect has recently been studied in psychophysics where the frequency or prevalence of a road hazard was found to have a strong effect on the chances of their successful detection [17]. It has also been observed in practice with some Level 2 systems, in which drivers tend to overtrust the system and begin to neglect looking at the road in favor of looking elsewhere, potentially engaging in distracting secondary activities [18]. In contrast, driver monitoring is becoming an important part of systems such as General Motors' Super Cruise, to help ensure that during highway Level 2 driving, humans continue to pay attention to the driving scene and are prepared to initiate or handle a takeover when required.

For all these reasons, driver monitoring systems are expected to become a more widespread safety feature in vehicles over the next few years. For example, in Europe, the New Car Assessment Program (NCAP) will require new cars to be fitted with DMS from 2023 to help reduce distracted and drowsy driving [19]. In the US, as part of the bipartisan infrastructure bill passed in 2021, auto manufacturers may be required to install DMS into new cars as early as 2026 to help prevent intoxicated drivers [20].

Despite the potential value of driver monitoring and the early-stage commitment from some governments and automakers towards DMS, there remain numerous challenges which the technology faces to maximize its positive impact over the next decade. For the remainder of this section, we examine two of the key challenges and describe efforts which we are pursuing to help tackle them.

2.1 Two Key Challenges

Separation of Data and Model in Machine Learning for Driver Monitoring
Modern supervised machine learning techniques, of the type used to train cutting edge driver monitoring systems, are data hungry. They require large image, video, and multi-modal (e.g. video plus CANbus) datasets, labelled with various metadata to capture the tasks of interest. Such labels range from relatively low-level tasks such as facial keypoint detection or eye feature extraction, to higher-level tasks such as drowsiness or alertness monitoring. Training dataset diversity is an important ingredient for run-time performance. It is often the case that, after deployment, driver monitoring systems exhibit certain failure modes, often as a consequence of lacking a particular mode of data in the training set. In such circumstances, it is common for DMS developers to add correctly annotated failure data to their training datasets, to allow their models to learn to behave correctly in those situations. In the ultimate limit of data and compute, this kind of training dataset augmentation can theoretically yield perfectly performing models.

Current practice, however, falls far short of this limit. While systems are being constantly improved and iterated upon, the tight feedback loop between data and model development, in which failure data can be flagged, annotated, and fed back into model training, is very difficult to set up. One reason for this is the organizational separation of field data, gathered extensively through system use by the automaker, and model, developed by a third party supplier, in the typical supply chain. Another is privacy: the ideal DMS is a closed-loop system, meaning all video data stays onboard the vehicle so as not to pose any privacy concerns for users. This means that data may not be straightforwardly harvested from fleets of users for development. A third challenge is the automated identification of failure modes. Since most systems cannot reliably estimate uncertainty, it is difficult to determine and sample data to improve the grey areas of system performance.

Integration of Driver Monitoring with Other ADAS Subsystems
The integration of DMS into other ADAS subsystems, if done well, could yield significant performance improvements, since ADAS behaviors can be modulated to better suit real-time driver states. For example, an automatic emergency braking sequence could initiate much earlier if the driver is known to be unaware of a potential risk, since the driver's acceptance of a warning or intervention will likely be significantly stronger. Similarly, lane-keeping assistance or blind-spot detection might behave more sensitively and warn a driver or augment the driver's controls earlier if the vehicle detects the driver is distracted or drowsy.

However, performing this integration well presents another key challenge. Safety systems which depend on both driver and driving state are by their nature harder to develop: there are likely to be more input edge cases compared to a non-integrated driver state system, because their input dimensionality is higher. Moreover, the types of situations in which these systems are targeted are likely to be more subtle. While today's AEB systems typically aim to deploy in the final one or two seconds before a likely impact, driver state-aware AEB systems may deploy several seconds earlier, expanding the operational design domain. As such, hand designing logical rules and processes to effectively govern the behavior of such integrated systems can be very challenging, if not impossible. To create integrated systems which work well, we believe that we will have to rely heavily on machine learning from data.

2.2 Two Ongoing Efforts

Prior research from our group has shown the value of using driver-facing video to infer various aspects of driver state, from elevated alertness [21] and attentional awareness [22], to heart rate estimation [23], 3D gaze estimation [24] and timing interactions between cars and drivers [25]. We briefly describe two additional ongoing efforts to address the challenges of developing driver monitoring for more complex estimation tasks.

DriverNet
To avoid the separation of data and model, we have developed an efficient multi-task neural network, dubbed DriverNet. The model, summarized in Fig. 2, takes driver-facing imagery as input and estimates a number of useful factors of driver state. These include 2D facial landmark tracking, coarse gaze estimation, eye openness estimation and various driver and image attributes. The model is trained from scratch in a supervised fashion using a weighted multi-task objective, minimizing the errors of various target outputs such as landmark estimation accuracy and gaze classification accuracy, using internally acquired datasets. There are several key advantages of this approach.

Firstly, the approach is flexible, and can be easily tuned towards different driver-facing camera positions, towards different desired outputs, or towards different computational budgets.

Secondly, the model is compute- and memory- efficient. Intermediate features are reused for various output tasks, and new tasks can reuse these features when helpful, leveraging the benefits of multi-task learning [26, 27]. For example, to augment the system with a face mask detector, an engineer can simply append the training dataset with images of masked drivers, and add an additional mask detector output to the head feature representation. By fine-tuning the head feature representation in this way, the overall system does not grow excessively in complexity as more tasks are added. Individual modules can also be tested relatively easily and efficiently.

Thirdly, by keeping data and model together, the overall system can be continuously, transparently improved, in the Toyota spirit of "*kaizen*". One mechanism to achieve system improvements is to sample data from a large driving fleet. The model can use proxy measures of uncertainty to identify challenging data points, which can be harvested, annotated and added to the training dataset. These proxy measures can be both direct (e.g. via image quality estimation outputs) and indirect (e.g. by detecting low likelihood predictions or inconsistencies between different outputs).

Contextual Driver Attention Estimation
The task of driving safely requires varying levels of visual attention from a driver, depending on a large number of factors such as speed, traffic density and intent, and road conditions. In scenarios where the time to impact with a real or potential hazard is short, such as driving through a narrow passage with many occlusions, the highest levels of attention are demanded: momentary glances away from the roadway significantly reduce the driver's ability to respond correctly to possible emerging hazards. Visual attention

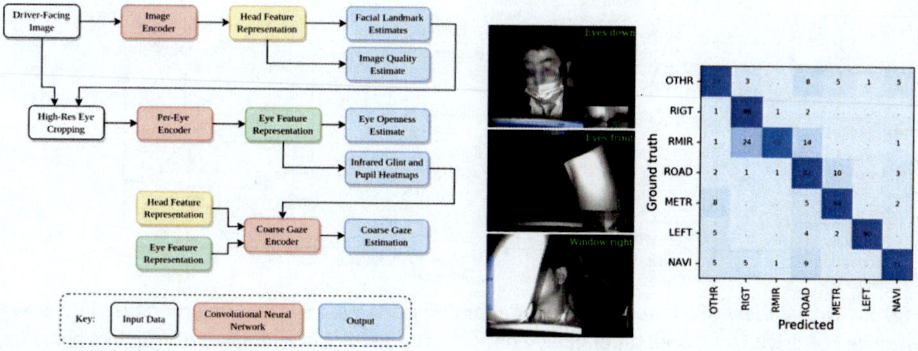

Fig. 2. *Left:* Overview of DriverNet model architecture. *Center:* Examples of DriverNet correctly estimating driver coarse gaze on challenging inputs with masks and glasses, poor illumination and occluded face views. Faces are blurred for anonymity. *Right:* Confusion matrix for coarse gaze estimation on a challenging test dataset, discriminating where the driver is looking among a discrete set of landmarks within the driving cabin, such as the road ahead (ROAD) or navigation console (NAVI).

should also be well-directed: in recent work we have shown that cuing visual attention to the wrong part of a driving scene at the wrong moment can significantly reduce hazard awareness [28]. As such, it may be valuable for a driver monitoring system to estimate not just where a driver is fixating in a given moment (gaze estimation), but how much attention they may have paid to different parts of the driving scene in recent history, both directly and peripherally. To perform this task of contextual driver attention estimation, a DMS may benefit from other sources of information beyond just driver-facing sensing. This is a step towards modeling the situational awareness of the driver [29, 30].

We are developing a contextual driver attention system, summarized in Fig. 3, which builds on top of the DriverNet model from the previous section. Our model combines information from a driver monitoring camera, as well as road-facing cameras and vehicle CANbus information, to estimate where a driver may be looking in the driving scene and what they may or may not be aware of.

Our goal is to develop a system which answers these questions by exploiting large datasets with rich models which are able to learn the correlations between behavior and outcomes across a range of driving conditions. Previous efforts to develop such contextual driver attention (and inattention) models were often heavily hand-engineered, resulting in systems which improved significantly upon non-contextual models but were restricted by various design limitations such as a large number of difficult-to-tune parameters (see e.g. [31, 32]). In contrast, with the potential abundance of multi-modal driving data available from a large vehicle fleet, we aim to create a data-driven rather than rule-driven system. We seek to leverage this data while acknowledging its characteristics – e.g. through a careful sampling of training samples towards examples of behaviors which lead to unfavorable outcomes (e.g. sudden braking or steering events). An early, incomplete version of our system was presented in [22], using passively observed in-lab driving gaze data without CANbus information.

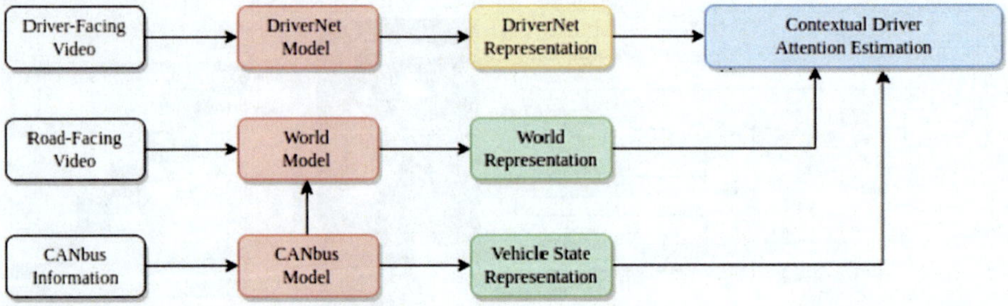

Fig. 3. An overview of a contextual driver attention estimation system, incorporating DriverNet estimates of driver state with an understanding of the road scene and the ego-vehicle's behavior within it.

2.3 Future Trends in Driver Monitoring

As we pursue research towards the better integration of driver state estimation within the HCID system, we note several important trends which will likely assist our approach. On the machine learning side, self-supervised/few-shot learning and "sim2real" (learning models with simulated data before transferring them to the real world) have begun to show great promise to improve the efficiency and performance of our models. For example, unsupervised gaze estimation methods such as the cross-encoder [33] can be used to pre-train the gaze encoder in DriverNet, significantly reducing the reliance on expensive annotated training data. Similarly, contrastive learning techniques may be used to learn good low-dimensional vehicle state and world representations for contextual driver attention estimation without the need for labels [34, 35]. In sim2real, diverse and near-photorealistic synthetic face datasets such as [36] will soon become more commonplace, replacing much of the need for acquiring real data, as well as providing a means of equitable evaluation of certain driver-facing models.

Beyond this, we are excited about the potential to harness the use of data sampled intelligently from large-scale driving fleets to efficiently learn our models and verify them "in the wild". We are also excited by the potential to combine driver state monitoring with information from wearable physiological sensors and other input modalities such as speech, as we work towards better personalization of the driving experience to individuals with all manner of different needs. Making these steps while respecting the privacy of individual users may require further developments in topics such as federated learning and differential privacy, but the potential personal and societal benefits of success make the research objective highly attractive.

3 Emergency Interventions: The Need for Expert Driving Skills

An important feature of the HCID system is emergency intervention: amplifying the driver by augmenting their inputs in circumstances where it deems that it can help provide better outcomes. In these situations, encoding expert driving skills into the system will help ensure that the vehicle handles these challenging situations successfully. Unlike the piecewise components that make up the ADAS constellation on vehicles today, this is envisioned as a single integrated system, as illustrated in Fig. 1. Thus, an intervention could occur when the system thinks it can either perceive, decide, or execute better than the human driver, whether due to the dynamic state of the vehicle, the specific scene, driver inattention, or a combination of those factors. In any of these cases, the system provides inputs to the vehicle from a single policy, a unified planning/control approach which could provide better outcomes when expanding the scope of ADAS across the narrow responsibilities of individual systems. This is especially true in complex, blended situations: in a brake-and-swerve maneuver, for example, the complex interactions between AEB, ABS, and ESC may provide suboptimal results. However, while subsuming the role of these disparate modules into one unified system could potentially offer great benefit, it also mandates a high level of responsibility to account for the critical vehicle dynamics that these proven production modules account for; it is against this backdrop that we discuss some of the key challenges and present research efforts in the area of autonomous vehicle control up to – and beyond – the limits of handling.

3.1 Key Benefits, Challenges and Present Research

There are fundamental limits to the actions an interventional system can take – they must lie within the possibilities governed by vehicle dynamics, available tire-road friction, and actuator saturation. The ability to confidently operate close to these physical limits widens the range of maneuvers available to the system. This not only increases the variety and severity of situations in which the system could intervene, but could also allow the system to intervene later during each scenario, thereby helping to improve the critical trade-off point between human and machine.

Automatic safety systems that modulate human inputs in response to critical dynamic states of the vehicle are already commonplace today: Electronic Stability Control (ESC) and Anti-Lock Braking systems (ABS) are federally mandated [37] in the United States, and have been shown to be effective in broad studies [38–40]. One central function of these systems is to prevent the vehicle from becoming open-loop unstable through rear tire saturation [41]. An illuminating method of viewing this task in vehicle dynamics terms is on the sideslip-yaw rate phase plane, wherein the rear tire saturation limit and maximum steady-state yaw rate can be viewed as a trapezoid, described by Beal et al. as the Stable Handing Envelope [42]. Although the implementation of these systems varies amongst manufacturers and models, the general principle can be broadly interpreted as trying to constrain the vehicle within a similar envelope.

Indeed, viewing the concept in these terms clarifies that this approach is fundamentally a trade-off: we sacrifice agility, manifested as motions in the phase plane at high yaw rate and sideslip, for stability. The proven effectiveness of these systems over many years of commercial implementation suggest that this trade-off is worthwhile in aggregate for the 'average' driver [38–40]. Yet, expert drivers (e.g. motorsports professionals) demonstrate the skill to operate in these unstable regions [43], and do so while balancing the costs and benefits of this increased agility with other objectives, e.g. staying within the road bounds and minimizing time. This capability to consider multiple, sometimes conflicting, objectives contrasts with current-generation ESC/ABS systems that, due to their limited sensing and actuation, can solely focus on preserving desired open-loop yaw dynamics, but not, for example, about the subsequent path of the vehicle in its environment. The advent of advanced automation with additional sensing and actuation modalities, however, suggests that expert driving systems of the near-future might not have to make this same simplistic trade-off. Instead, they could use these open-loop unstable dynamics as necessary to avoid accidents, as part of a unified approach to safe navigation – this is the vision that the HCID concept embodies.

For this approach, a particularly relevant expert driver example is the motorsports discipline of 'drifting', wherein drivers purposefully saturate the rear tires of a vehicle, and compete to achieve high sideslip while minimizing the distance to, but not colliding with, track boundaries and other obstacles. This impressive display of skill has also been explored in the literature. Early works showed that drifting is an unstable equilibrium that can be seen in various levels of vehicle fidelity [44]–[46]. Several control approaches to stabilize the dynamic states of the vehicle around a single equilibrium have been demonstrated, including Linear Quadratic Regulator (LQR) control [43], dynamic surface control [47], and reinforcement learning [48]. When combining vehicle state stabilization with the task of tracking a path, there are more states than inputs, and the problem becomes underactuated. One feedback control approach used simplifying assumptions to decompose the problem of drifting in a steady-state circle into tracking vehicle course with the steering actuator, and sideslip with discrete bang-bang engine torque pulses [49]. Another approach tracked vehicle position and sideslip with nonlinear model inversion, and left vehicle velocity as an uncontrolled, but stable, zero dynamic [50]. This approach was experimentally shown to generalize to both slowly changing drifting equilibriums, and rapid transitions between them [51], and emphasized the importance of accounting for wheelspeed dynamics when operating at high tire slip.

These works demonstrate the importance of modelling the appropriate dynamics, and the trade-offs inherent when controlling underactuated systems, as applied to the specific task of autonomous drifting. One promising technique for propagating these insights to more general tasks is Nonlinear Model Predictive Control (NMPC). In NMPC, the cost function and constraints can be general nonlinear functions, allowing for detailed model fidelity, and the encoding of elaborate desired behaviors. The complexity of these representations, however, needs to be balanced against convergence times and horizon length. Recently, several works have demonstrated the viability of this approach for limit handling situations. By cascading several levels of decreasing model complexity along the planning horizon, Laurense et al. demonstrated an NMPC approach for the task of autonomous racing that balances near-term fidelity with longer horizon lengths [52]. A single-track representation with longitudinal weight transfer and non-linear, force-coupled tires was used as the short-range, highest fidelity model, and captured the dynamics well when approaching the limits of tire saturation. By combining a similar model with a computationally efficient representation of the environment and ego vehicle in an NMPC framework, Brown et al. represented a scenario with a sudden vehicle cut-in [53]. In experimental testing, the controller was able to negotiate this emergency double lane change scenario in a combined braking and steering maneuver that deftly co-ordinated lateral-longitudinal force coupling close to the friction limits. Other works studying dynamic autonomous driving have leveraged the flexibility of NMPC to explore varied concepts, including incorporating measures of uncertainty [54, 55], operating in unstructured environments [56], shared control [57], and ensuring feasible contingency plans [58].

3.2 Ongoing Research Efforts

Indeed, taken together, these recent works suggest that combining the vehicle dynamics and control design insights from the drifting-specific approaches with the generalized framework of NMPC could yield a control scheme that extends its operational domain *beyond* the point of tire saturation, and is capable of deftly driving both inside and outside the open-loop stability bounds. Recently, Toyota Research Institute has developed a step towards such a unified approach, and experimentally demonstrated an NMPC controller that can smoothly transition from dynamic, non-equilibrium drifting to grip driving, while accounting for multiple objectives including road bounds (Fig. 4(b)).

This experimental result is enabled by the incorporation of a few novel features. Firstly, the driven rear axle wheelspeed is included as an explicit vehicle state, building on work that emphasized the importance of compensating for wheelspeed dynamics when tires are operating in a high positive slip regime [50]. This wheelspeed is then used to compute longitudinal slip, which is subsequently included in an isotropic coupled-slip model for the rear tires that smoothly transitions between the grip and sliding regimes. Simple first-order models are used to account for engine torque and steering actuator dynamics, and engine torque/steering angle slew rates are then treated as inputs to the vehicle model with hard upper/lower bounds. Secondly, to compensate for the finite length of the control horizon, a cost is added to represent sideslip stability at the terminal stage. This is formulated as a 2-norm penalty on deviating from a first-order stable sideslip dynamic (e.g. $\dot{\beta} = -k_\beta e_\beta$), where the time constant k_β is a tuning parameter. This is inspired by the derivation of the controllability envelopes discussed in [51]. Other costs in the objective function include a quadratic cost on tracking the sideslip prescribed by a reference trajectory, quadratic costs on input slew rate deviations relative to the reference trajectory, and a large quadratic penalty on exceeding prescribed maximum/minimum bounds on lateral error that represent the available road area.

With any NMPC formulation, the complexity of the model and number of included states/inputs has to be balanced against convergence times. This is particularly important for a system that is expected to intervene after an edge case event – for example, spinning after hitting a patch of ice – which can be both very rapid and stochastic. To evaluate such a hand-off in our experimental setup, the NMPC controller is started after a separate classical approach suddenly destabilizes the vehicle by locking the rear axle with strong braking pressure. In Fig. 4(a), which shows the reference path and road bounds, this segment is indicated by the orange line. After this initiation phase, the NMPC controller takes over, up to and including a return to straight-ahead grip driving at the end of the test trajectory; this section is indicated by the green line.

This approach was tested on the 'Keisuke' testbed, a Toyota Supra that has been specially customized for autonomous driving research. It is equipped with computer-controlled steering, throttle, clutch displacement, sequential transmission, and individual wheel braking. Vehicle state information is obtained from a dual-antenna Oxford Technical Systems RT4003 RTK-GNSS-aided INS system at a rate of 250 Hz, and the NMPC controller runs on an x86 computer. For the purposes of data collection with expert drivers in a controlled environment, the suspension, engine, transmission, chassis, and safety systems (e.g. rollcage, fire suppression) have been modified to be similar to that used in Formula Drift competitions.

Experiments were conducted at Thunderhill Raceway, on the 2-mile 'West' track. A video showcasing a few experimental runs can be found at https://www.youtube.com/watch?v=MfU5_gzqPaM. In (Fig. 4(b)), a composite photo superimposes stills from an overhead video of a test at 0.5s intervals, showing the path and pose of the car throughout the closed-loop trajectory. Overall, performance was robust over several back-to-back runs, with mean solve times around ~0.02 s The non-linear program (NLP) was formulated with the aid of the open-source CasADi toolbox [59], and solved online with the open-source IPOPT interior-point optimization package [60].

An insightful way to view the data is in the sideslip-yaw rate phase plane, as depicted in Fig. 5. The reference trajectory, depicted as black dashed lines, shows smooth motions in the phase plane without dwelling, indicating its dynamic, non-equilibrium characteristic. Also shown are the stable handling envelopes, as defined by Beal et al., at the maximum and minimum speeds of the trajectory. The measured data starts out in the grip driving stage (blue) within these envelopes, then is rapidly taken outside these bounds during the handbrake destabilization stage (orange). Thereafter, the NMPC controller (green) guides the vehicle throughout the phase plane, smoothly changing directions while fully exploiting states with high yaw rate and sideslip – and simultaneously keeping the car within the bounds of the road. It then takes the vehicle back to grip driving within the stable region, after which the experiment is ended and the safety driver takes over control.

Indeed, the span of the phase trajectory, when compared to the significantly smaller size of the stable region, makes clear the benefits of this approach: the controller can access much greater yaw rates and sideslips, and this increased agility could prove beneficial in emergency situations. This capability demonstration, directed towards the motorsports-inspired challenge problem of drifting, is a step towards the goal of a generalized controller that can utilize high tire saturation and open-loop unstable dynamics. In the next stage of development, we hope to explore situations in which an NMPC formulation uses these unstable regions as needed, that is, where high sideslip maneuvers are an emergent, rather than explicitly targeted, behavior.

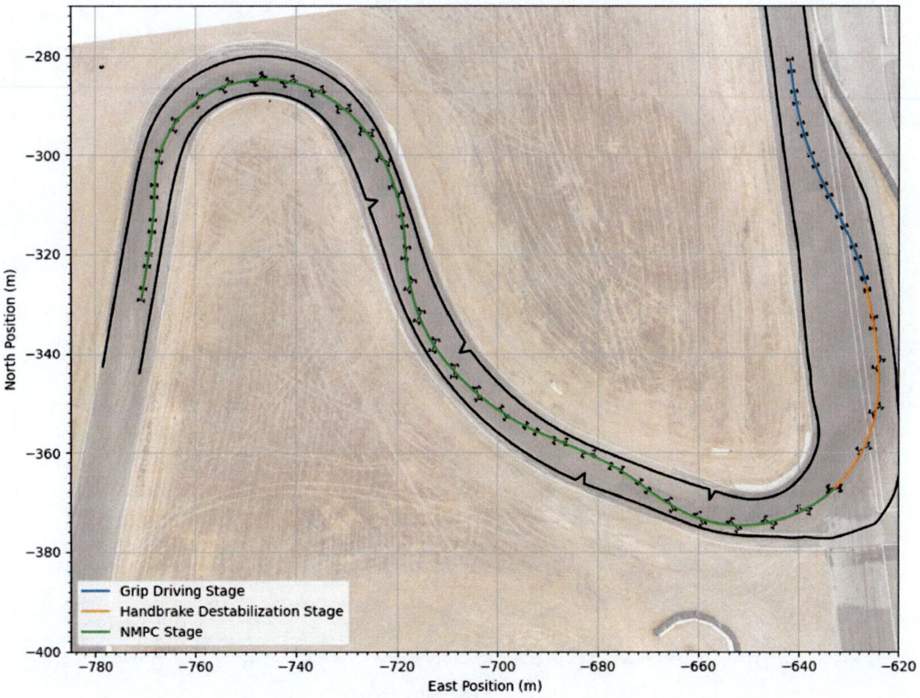

(a)

Fig. 4(a). The reference path for the grip driving (blue), handbrake destabilization (orange), and NMPC (green) stages. The pose of the vehicle is overlaid at several points throughout the path. Also shown are the road bounds (black) used in the online NMPC solver.

(b)

Fig. 4(b). Stills from an overhead video of an experimental test superimposed at 0.5 s intervals, showing the path and pose of the car throughout the closed-loop trajectory.

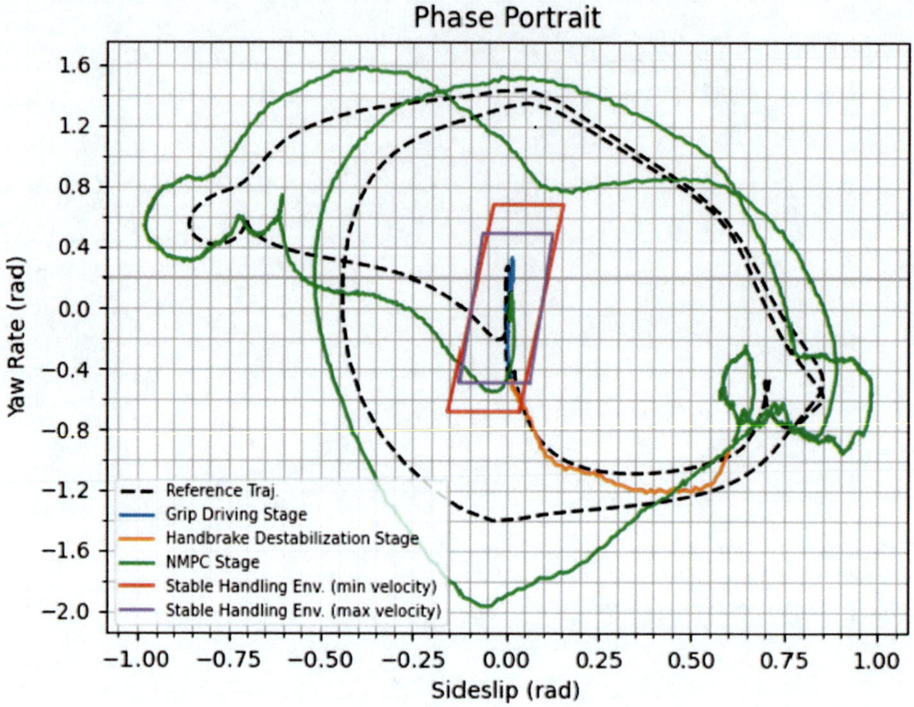

Fig. 5. Experimental data depicted on the yaw rate vs. sideslip phase plane, segmented into the grip driving stage (blue), handbrake destabilization stage (orange), and NMPC stage (green). Also shown are the stable handle envelopes [42] at the maximum (purple) and minimum (red) velocities, as well as the reference trajectory (black dashed).

4 Scene Understanding: Risk Modeling and Behavior Forecasting

An HCID system that protects a driver requires an understanding of how the scene around the vehicle will evolve, and how the driver may react to it. In order to intervene successfully, the system needs to know that its intervention policy provides better safety than what the human driver will do. Thus, an HCID system should strive to accurately gauge scene risk, as well as to understand road users' intent and predict their behaviors.

4.1 Scene Risk and Uncertainty Estimation

An important component of the HCID system is the ability to evaluate the world around the vehicle, and foresee possible risks and outcomes. Risky events occur due to many factors [61], and may be reasoned about from different perspectives. The many perspectives of risk lead to different technical approaches for safety (consider PCS, DMS and other safety systems).

A common approach for risk considers possible collisions. This approach lends itself easily into efficient planning [62], warning frameworks, and verification approaches. Another important aspect of risk is reasoning about unseen obstacles and occlusions.

Several approaches can mitigate risks from imperfect perception – e.g. using cues from other drivers [63], via non-line-of-sight (NLOS) computer vision [64], and modelling of risk due to occlusions [65]. Other risk factors such as decision-making errors lead to different warning and intervention approaches. While detecting human errors is not trivial, imperfect estimates of deviation from the usual policy can be used to detect risks from both the driver and the environment. Another cause of risk regards drivers and interaction with the scene: the mismatch between ego- and ado- situational awareness or interpretation of the scene. Drivers' situational awareness can be improved by warning systems, and awareness can be estimated better, as described in Sect. 2. Gauging ado-agents' awareness is another important part of defensive driving. For intelligent vehicles to make decisions, they depend also on the car's imperfect situational awareness - a major challenge for Automated Driving (AD) deployment. While planning and prediction approaches can be made more robust to perception failures [66, 67], the safety-critical nature of driving leaves room for improvement in this domain.

Finally, reasoning about risk extends into the automotive ecosystem. Insurance companies instrument vehicles to check for the presence of "high-G" events that may indicate evasive or aggressive maneuvers. Road authorities and vehicle suppliers are looking to add vehicle-to-vehicle and vehicle-to-infrastructure systems to provide additional warning layers [68]. Such actors may need to be taken into account.

More broadly, there are multiple sources of uncertainty in driving, due to perception limitations and the nature of latent human intent, but also the uncertainty inherent to the nature of human behavior. We account for many of them in our approach, including the influence of perception errors on the vehicle behavior [66], influence of occlusion limitations [65], and the inherent uncertainty of human actions [69–72]. Even estimating a single optimal plan may not be trivial – humans have their own perception of risk and are reasonable at it when they are alert. An AI deviating from such norms may cause other road users to make mistakes.

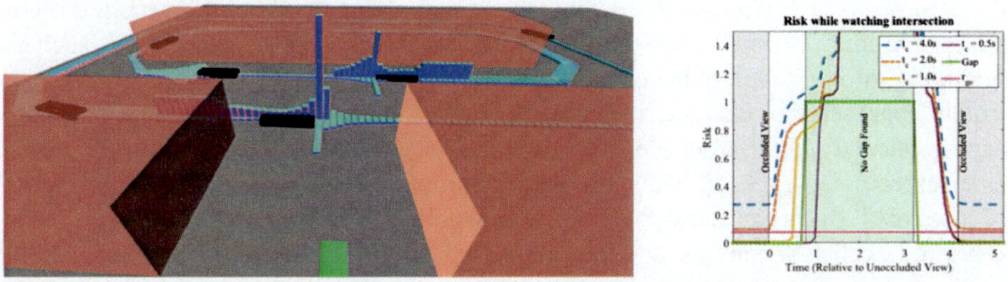

Fig. 6. Intersection risk modeling done in our group. Left: intersection risk visualization, demonstrating occluding building, and occupancy and alertness estimates for incoming vehicles. Right: illustration of computed quantities in the model for merge gap estimates given occluded areas and different intersection clear times(t_c).

Research in our group addresses many of these perspectives of risk. It includes both mitigating occlusion via computer vision NLOS approaches [64] as well as models that better estimate the risk inherent in occlusions and take into account the agents in the scene a model of their inattention, as we have shown in our recent model [65], shown in Fig. 6. This model demonstrates novel representations that capture "just the right amount" of information. It takes multiple factors into account and captures occlusion information, driver distraction reasoning and approximate velocities into a risk model that affords both warning and parallel autonomy.

Another important aspect researched in our team involves the human aspect of risk – e.g. in better models to understand what the driver looks at when we reason about risk [22], estimate their behavior and the factors that underlie it [69–71, 73, 74], as well as how that behavior impacts safety approaches and future optimal actions for the HCID system. We expect major gains to be achieved via better approaches to fuse these different perspectives and cues, and are working towards more holistic approaches.

4.2 Intent Estimation and Prediction

Road agent prediction estimates future actions of vehicles and agents in the environment – both the driver and those external to the vehicle. Prediction plays a role in several aspects of HCID systems, beyond AD stacks' requirements. The behavior estimates inform HCID systems of the driver's intent in order to gauge future risk, as well as for driver-system collaboration. Prediction also considers how other agents would react to what the driver or intelligent vehicle will be doing, e.g., for planning with a better model of how the world evolves, or for verification and validation [75, 76]. Unlike AD stacks, Guardian systems have a narrower ability to plan around prediction limitations.

Prediction commonly leverages several representations, including trajectory, maneuvers, and intents [77, 78]. Multi-agent trajectory prediction is often defined as the following problem: Given the positions of N_{AGENTS} agents for the last T_{PAST} timesteps, predict their state for the next T_{FUTURE} timesteps. Other formulations of the problem are area-centric (Eulerian) approaches that model spatial area risk and occupancy [65, 79, 80]. These are useful for planning and risk estimation, and are sometimes combined with the individual agents' predictions. Many of the recent approaches form an encoder-decoder architecture. The latent factors underlying these behavior models often address temporal and spatial reasoning, as well as other types of structure priors such as underlying rules [81, 82], maneuvers [70, 83], agents' reward and goal models [84, 85], or other approaches to characterize human behavior in the world [74]. Predictor inputs vary significantly in different scenarios – from raw sensors [69, 86] to completely processed and curated tracks [87–89]. An important input comes from maps, represented as polynomials [70], raster maps [84], or graph neural networks [90, 91]. However, maps go beyond explicit semantics, and locale-specific customs need to be captured as well. Another set of important inputs relates to humans with whom the car interacts, such as the subtle visual cues of pedestrians [92–95] and driver-facing cues when estimating future actions of the driver [96]. Driver awareness of the scene [97] is especially important, as it both hints what a driver may intend to do and models the relevant scene information based on what the driver will react to. The predicted outcomes are eventually used by a downstream task, e.g., a planner or a warning system. The interplay between prediction and planning has been addressed in both communities' literature. Planning literature has focused on leveraging predictions into planning approaches [98] or even jointly planning and predicting [99, 100]. Prediction literature has examined how downstream tasks affect quality measures [71, 73, 101].

Some approaches combine the latent structure and the numerics of the predictor to allow more efficient sampling that is better suited to the task at hand [70, 72, 73]. This is important due to both the ability to handle rare, but risky events [102, 103], and also the ability to mimic humans' intuitive ability to conjure and focus on relevant scenarios amidst the full set of possible outcomes, and use these to plan and explain plans.

Current challenges and trends are numerous. In addition to overall improvements in the different components, current efforts in prediction and behavior modeling include:

- Extending the prediction horizon to ~10 s [87] or more [89, 104], affecting both the models and the metrics. Especially important is good coverage of as many occurrences as possible [70, 105, 106].
- Scenarios where additional cues are needed, e.g., considering pedestrian forecasting as a separate prediction effort [92, 94, 95], or using driver-facing cameras in ego-intent prediction.
- Modeling agent interactions in a realistic and data-efficient way, since predictive approaches still lack the innate ability of humans to reason about complex interactions without a full roll-out of every scenario. This is important, too, for explainability, itself a key component to interaction with humans for both safety verification and communication with other human road users.
- Handling of rare events, which challenges the data collection, representation, and evaluation of predictors. While datasets continue to expand and include more interesting events, evaluations and representations still focus on high-probability events, which may be insufficient for safety validation.
- Handling different semantics such as rules, maneuvers, and patterns is important in how we reason about road behaviors. Rules, however, are often broken (consider how drivers observe speed limits), and this tendency must be addressed in prediction. Such common-sense use of rules and patterns goes beyond formal specifications, and becomes important to an HCID system.
- The role of prediction in downstream tasks for an HCID system includes novel metrics of success, such as explainability and completeness, in order to be a true instance of human robot interaction.

Our team has been addressing many questions arising from prediction and behavior modeling for an HCID system. The use of latent context maps [107] allows us to reduce the dependency on perfect high-definition maps, while providing robust performance where maps are not annotated. At the same time, latent maps can encode location-specific information and behaviors that may be learned by observing how other agents act in each place.

Some of our work explores how the underlying structure of human behavior is reflected in prediction. This includes rules, but unlike the use of rules in planning, here we must acknowledge that people may only loosely follow rules. Using rules as an inductive bias [82, 108], we can learn behavior with less data, even when the data and predictions include examples that break the rules [81]. Other approaches we have taken leverage hybrid control [72] and predict maneuver sequences in parallel to the trajectories, improving accuracy at longer horizons and yet allowing predictions that do not adhere closely to maneuvers. Finally, partial supervision from labels [70], and even language descriptions [74], are used towards more robust and explainable predictions.

Ego-driver prediction brings its own set of considerations, both in terms of additional inputs, and the estimation of confidence in the prediction. As our recent work shows [69], leveraging additional information from the vehicle affords a significant contribution to prediction accuracy. The same work also demonstrates the importance and feasibility of regressing the predictor's confidence: we demonstrate how such a confidence estimate allows us to improve prediction via predictor fusion and to avoid usage of sub-optimal predictors for downstream tasks. This regression of confidence is later shown to be crucial when the regression is made to be task-specific, and we demonstrate how task-specific uncertainty estimates can be used, e.g., to improve performance in a collision avoidance system [71]. As shown in Fig. 7, even approximate estimates on the predictor performance can be extremely useful if they are better tailored to the role of prediction within the tasks that an HCID system is set to accomplish. We have further developed this approach towards a more complete approach for task-informed prediction [73] that shapes the prediction samples towards improved decision making.

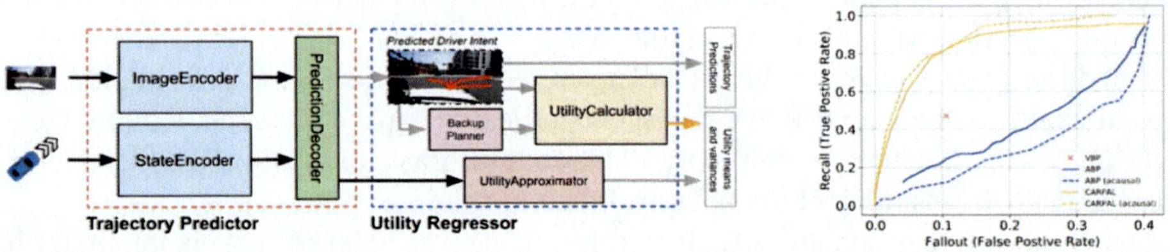

Fig. 7. In the CARPAL [71] approach, we predict future behavior and approximate the downstream task effects. This allows us to improve results in downstream tasks such as collision warnings. Left: the system structure. Right: an example ROC curve for a warning system with our (CARPAL) utility estimation.

5 Next Steps: Human-Machine Interaction

The advancements described above in human state estimation, scene representation and prediction, and vehicle controller performance, i.e., the HCID system has created opportunities for new ways of interacting with the human driver to bring about successful outcomes. To that end, the policies of the HCID system take into consideration the joint representation of the environment, vehicle, and human driver to output coordinated information to the driver. This joint policy consists of phases of interaction with the user that increase in severity given the risk of the driving situation (i.e., ranging from impending to imminent collision). These phases include: (1) enhancing the driver's situation awareness, (2) suggesting recommended actions to the driver, (3) augmenting the driver's intended actions, (4) intervening to complete actions for the driver.

For example, extended prediction horizons based on probabilistic models will necessitate new ways of communicating system uncertainty to the human. This communication method may include novel auditory cues and visualizations of prediction and planning uncertainty through the use of spatio-temporal variations in color, saturation, shape, and symbols, and text. Also, enhanced driver state estimation coupled with widened maneuver availability of the vehicle controller allows the HCID concept system to offer a broader opportunity for blended control with the human driver. Haptic control in particular, can be used more widely with such a system as an unobtrusive, intuitive modality continuously guiding a human driver away from potential hazards while still preserving the general intended trajectory of the driver [109]. Each of these new HMI modalities will need to be designed to contribute to scaffolding the drivers' creation of an accurate mental model of how the system works. This mental model accuracy is critical to avoiding errors based on incorrect assumptions about system operation [110], which will become more complex as humans interact with increasingly automated vehicles.

To facilitate the creation of a holistic system that does not excessively distract the driver from the main driving task while achieving the desired outcomes, the design thinking process should be applied. Smaller, iterative user studies to evaluate the HMI prototypes should be performed to provide insights and learnings for new design iterations of the HMIs. Concomitantly, new subjective and objective performance measures and methods should be developed to evaluate multimodal driving HMIs involving shared control and AI.

6 Conclusion

The HCID concept as proposed by TRI leverages increasing automation of vehicles to improve overall driver and vehicle safety as well as performance. It aims to do this by enabling collaboration and cooperation between the AI and the human. HCID seeks to create rich joint representations from data from the vehicle, environment and human. Policies derived from these representations can, with the appropriate driver interactions, improve safety outcomes. This concept differs from traditional ADAS constellations by using all available data from the vehicle, environment and human to build representations that give the best understanding of the situation holistically instead of focusing on narrow use cases like AEB. Furthermore, by building policies off these joint representations, interactions with the driver can also be more cohesive and effective without the potentially confusing interaction of multiple separate ADAS features. The creation of these joint representations and downstream policies shares many similarities with fully automated driving, resulting in many common technical challenges. Some of the core technical challenges for HCID include understanding the driver state, building expert driving skills and creating effective interactions. This chapter has reviewed some of the latest TRI research in these areas including work in driver monitoring, limit handling control, risk estimation, modeling, and prediction of human behavior on the road.

Acknowledgements. The authors would like to acknowledge the work of their research colleagues at TRI who have contributed to many of these results. These include Alex Green, Cyrus Huang, Deepak Gopinath, Hanh Nguyen, Hiroshi Yasuda, John Gideon, John Subosits, Kazunori Nimura, Manuel Kuehner, Michael Thompson, Mitsuhiro Mabuchi, Paul Drews, Phung Nguyen, Tetsuya Ishikawa, Thomas Balch, Velin Dimitrov, and Xiongyi Cui. The authors would also like to express their appreciation to Ryan Eustice, Gill Pratt, and many other colleagues at TRI that have worked to advance the HCID and Toyota Guardian™ concepts.

References

1. Montgomery, W.D.: Public and private benefits of autonomous vehicles. Transp. Res. Board (2018). https://trid.trb.org/view/1522211
2. Bagloee, S.A., Tavana, M., Asadi, M., Oliver, T.: Autonomous vehicles: challenges, opportunities, and future implications for transportation policies. J. Mod. Transp. **24**(4), 284–303 (2016). https://doi.org/10.1007/s40534-016-0117-3
3. Koopman, P., Wagner, M.: Challenges in autonomous vehicle testing and validation. SAE Int. J. Transp. Saf. **4**(1), 15–24 (2016)
4. Barabás, I., Todoruţ, A., Cordoş, N., Molea, A.: Current challenges in autonomous driving. IOP Conf. Ser. Mater. Sci. Eng. **252**(1), 012096 (2017)
5. Leonard, J.J., Mindell, D.A., Stayton, E.L.: Autonomous vehicles, mobility, and employment policy: the roads ahead. MIT Task Force on Work of the Future, July 2020. https://workofthefuture.mit.edu/wp-content/uploads/2020/08/WotF-2020-Research-Brief-Leonard-Mindell-Stayton.pdf
6. Xing, Y., Huang, C., Lv, C.: Driver-automation collaboration for automated vehicles: a review of human-centered shared control. In: 2020 IEEE Intelligent Vehicles Symposium (IV), pp. 1964–1971, October 2020
7. Traffic Safety Facts: NHTSA, March 2018. https://crashstats.nhtsa.dot.gov/Api/Public/ViewPublication/812506
8. Eustice, R.: The road to vehicle automation: a Toyota Guardian approach. In: SPIE Future Sensing Technologies, vol. 11525, p. 1152502, November 2020
9. Hof, R.: Toyota: 'Guardian Angel'cars will beat self-driving cars. Forbes (2016)
10. Grace, R., et al.: A drowsy driver detection system for heavy vehicles. In: 17th DASC. AIAA/IEEE/SAE. Digital Avionics Systems Conference. Proceedings (Cat. No.98CH36267), vol. 2, pp. I36/1–I36/8, October 1998
11. Smith, P., Shah, M., da Vitoria Lobo, N.: Monitoring head/eye motion for driver alertness with one camera. In: Proceedings 15th International Conference on Pattern Recognition. ICPR-2000, vol. 4, pp. 636–642, September 2000
12. Fletcher, L., Loy, G., Barnes, N., Zelinsky, A.: Correlating driver gaze with the road scene for driver assistance systems. Rob. Auton. Syst. **52**(1), 71–84 (2005)
13. Manstetten, D., et al.: The evolution of driver monitoring systems: a shortened story on past, current and future approaches how cars acquire knowledge about the driver's state. In: 22nd International Conference on Human-Computer Interaction with Mobile Devices and Services, Oldenburg, Germany, pp. 1–6, October 2020
14. LeCun, Y., Bengio, Y., Hinton, G.: Deep learning. Nature **521**(7553), 436–444 (2015)
15. Khan, M.Q., Lee, S.: Gaze and eye tracking: techniques and applications in ADAS. Sensors **19**(24) (2019). https://doi.org/10.3390/s19245540
16. Greenlee, E.T., DeLucia, P.R., Newton, D.C.: Driver vigilance in automated vehicles: hazard detection failures are a matter of time. Hum. Factors **60**(4), 465–476 (2018)
17. Kanda, S., Kosovicheva, A., Wolfe, B., Wolfe, J.M.: Prevalence effects on the road: Rare hazards are often missed. J. Vis. **21**(9), 2968 (2021)
18. Morando, A., Gershon, P., Mehler, B., Reimer, B.: A model for naturalistic glance behavior around Tesla Autopilot disengagements. Accid. Anal. Prev. **161**, 106348 (2021)
19. Fredriksson, R., Lenné, M.G., van Montfort, S., Grover, C.: European NCAP program developments to address driver distraction, drowsiness and sudden sickness. Frontiers Neuroergonomics **2**, 33 (2021)
20. Elliott, P.: Buried in the bipartisan infrastructure bill: in-car breathalyzers. Time, 03 August 2021. Accessed 10 Jan 2022. https://time.com/6086981/bipartisan-infrastructure-bill-breathalyzers/

21. Gideon, J., Stent, S., Fletcher, L.: A multi-camera deep neural network for detecting elevated alertness in drivers. Presented at the IEEE International Conference on Acoustics, Speech and Signal Processing (ICASSP), April 2018
22. Gopinath, D., et al.: MAAD: a model and dataset for 'attended awareness' in driving. Presented at the IEEE/CVF International Conference on Computer Vision Workshops (ICCVW) (2021)
23. Gideon, J., Stent, S.: Estimating heart rate from unlabelled video. Presented at the IEEE/CVF International Conference on Computer Vision Workshops (ICCVW), Montreal, BC, Canada (2021)
24. Kellnhofer, P., Recasens, A., Stent, S., Matusik, W., Torralba, A.: Gaze360: physically unconstrained gaze estimation in the wild. Presented at the IEEE/CVF International Conference on Computer Vision (2019)
25. Wu, T., Martelaro, N., Stent, S., Ortiz, J., Ju, W.: Learning when agents can talk to drivers using the INAGT dataset and multisensor fusion. Proc. ACM Interact. Mob. Wearable Ubiquit. Technol. **5**(3), 1–28 (2021)
26. Kokkinos, I.: UberNet: training a 'Universal' convolutional neural network for low-, mid-, and high-level vision using diverse datasets and limited memory. Presented at the IEEE/CVF Conference on Computer Vision and Pattern Recognition (2017)
27. Crawshaw, M.: Multi-task learning with deep neural networks: a survey. arXiv [cs.LG], 10 September 2020. http://arxiv.org/abs/2009.09796
28. Wolfe, B., Kosovicheva, A., Stent, S., Rosenholtz, R.: Effects of temporal and spatiotemporal cues on detection of dynamic road hazards. Cogn. Res. Princ. Implic. **6**(1), 1–15 (2021). https://doi.org/10.1186/s41235-021-00348-4
29. Endsley, M.R.: Toward a theory of situation awareness in dynamic systems. Hum. Factors **37**(1), 32–64 (1995)
30. Wolfe, B., Sawyer, B.D., Rosenholtz, R.: Toward a theory of visual information acquisition in driving. Hum. Factors, 18720820939693 (2020)
31. Ahlström, C., Georgoulas, G., Kircher, K.: Towards a context-dependent multi-buffer driver distraction detection algorithm. IEEE Trans. Intell. Transp. Syst., 1–13 (2021)
32. Kircher, K., Ahlström, C.: Issues related to the driver distraction detection algorithm AttenD. Presented at the First International Conference on Driver Distraction and Inattention (2009)
33. Sun, Y., Zeng, J., Shan, S., Chen, X.: Cross-encoder for unsupervised gaze representation learning. Presented at the IEEE/CVF International Conference on Computer Vision (2021)
34. Pan, T., Song, Y., Yang, T., Jiang, W., Liu, W.: VideoMoCo: contrastive video representation learning with temporally adversarial examples. In: Proceedings of the IEEE/CVF Conference on Computer Vision and Pattern Recognition, pp. 11205–11214 (2021)
35. Qian, R., et al.: Spatiotemporal contrastive video representation learning. In: Proceedings of the IEEE/CVF Conference on Computer Vision and Pattern Recognition, pp. 6964–6974 (2021)
36. Wood, E., et al.: Fake it till you make it: face analysis in the wild using synthetic data alone. Presented at the IEEE/CVF International Conference on Computer Vision (2021)
37. Federal Motor Vehicle Safety Standards, 49 C.F.R § 571.126
38. Farmer, C.M.: Effects of electronic stability control: an update. Traffic Inj. Prev. **7**(4), 319–324 (2006)
39. Ferguson, S.A.: The effectiveness of electronic stability control in reducing real-world crashes: a literature review. Traffic Inj. Prev. **8**(4), 329–338 (2007)
40. Dang, J.N.: Statistical analysis of the effectiveness of electronic stability control (ESC) systems-final report (2007). https://trid.trb.org/view/838850
41. Liebemann, E.K., Meder, K., Schuh, J., Nenninger, G.: Safety and performance enhancement: the Bosch electronic stability control (ESP). SAE Technical Paper (2004). https://www.sae.org/gsdownload/?prodCd=2004-21-0060

42. Beal, C.E., Christian Gerdes, J.: Model predictive control for vehicle stabilization at the limits of handling. IEEE Trans. Control Syst. Technol. **21**(4), 1258–1269 (2013). https://doi.org/10.1109/tcst.2012.2200826
43. Velenis, E., Katzourakis, D., Frazzoli, E., Tsiotras, P., Happee, R.: Steady-state drifting stabilization of RWD vehicles. Control Eng. Pract. **19**(11), 1363–1376 (2011)
44. Voser, C., Hindiyeh, R.Y., Gerdes, J.C.: Analysis and control of high sideslip manoeuvres. Veh. Syst. Dyn. **48**(sup1), 317–336 (2010)
45. Ono, E., Hosoe, S., Tuan, H.D., Doi, S.: Bifurcation in vehicle dynamics and robust front wheel steering control. IEEE Trans. Control Syst. Technol. **6**(3), 412–420 (1998)
46. Hindiyeh, R.Y., Gerdes, J.C.: Equilibrium analysis of drifting vehicles for control design. In: ASME 2009 Dynamic Systems and Control Conference, pp. 181–188, September 2010
47. Hindiyeh, R.Y., Christian Gerdes, J.: A controller framework for autonomous drifting: design, stability, and experimental validation. J. Dyn. Syst. Meas. Control **136**(5) (2014). https://doi.org/10.1115/1.4027471
48. Cutler, M., How, J.P.: Autonomous drifting using simulation-aided reinforcement learning. In: 2016 IEEE International Conference on Robotics and Automation (ICRA), pp. 5442–5448, May 2016
49. Werling, M., Reinisch, P., Gröll, L.: Robust power-slide control for a production vehicle. Int. J. Veh. Auton. Syst. **13**(1), 27–42 (2015)
50. Goh, J.Y., Goel, T., Christian Gerdes, J.: Toward automated vehicle control beyond the stability limits: drifting along a general path. J. Dyn. Syst. Meas. Control **142**(2) (2020). https://doi.org/10.1115/1.4045320
51. Goh, J.Y.M.: Automated vehicle control beyond the stability limits. Ph.D., Stanford University (2019). https://searchworks.stanford.edu/view/13250106
52. Laurense, V.A., Gerdes, J.C.: Long-horizon vehicle motion planning and control through serially cascaded model complexity. IEEE Trans. Control Syst. Technol. **30**(1), 166–179 (2022)
53. Brown, M., Gerdes, J.C.: Coordinating tire forces to avoid obstacles using nonlinear model predictive control. IEEE Trans. Intell. Veh. **5**(1), 21–31 (2020)
54. Dallas, J., Wurts, J., Stein, J.L., Ersal, T.: Contingent nonlinear model predictive control for collision imminent steering in uncertain environments. IFAC-PapersOnLine **53**(2), 14330–14335 (2020)
55. Brown, M., Gerdes, J.C.: Robust stabilization and collision avoidance through minimizing open-loop velocity uncertainty. In: 2020 IEEE Intelligent Vehicles Symposium (IV), pp. 259–264, October 2020
56. Yoon, Y., Shin, J., Kim, H.J., Park, Y., Sastry, S.: Model-predictive active steering and obstacle avoidance for autonomous ground vehicles. Control Eng. Pract. **17**(7), 741–750 (2009)
57. Schwarting, W., Alonso-Mora, J., Paull, L., Karaman, S., Rus, D.: Safe nonlinear trajectory generation for parallel autonomy with a dynamic vehicle model. IEEE Trans. Intell. Transp. Syst. **19**(9), 2994–3008 (2018)
58. Alsterda, J.P., Brown, M., Gerdes, J.C.: Contingency model predictive control for automated vehicles. In: 2019 American Control Conference (ACC), pp. 717–722, July 2019
59. Andersson, J.A.E., Gillis, J., Horn, G., Rawlings, J.B., Diehl, M.: CasADi: a software framework for nonlinear optimization and optimal control. Math. Program. Comput. **11**(1), 1–36 (2018). https://doi.org/10.1007/s12532-018-0139-4
60. Wachter, A., Biegler, L.T.: On the implementation of a primal—dual interior point filter line search algorithm for large-scale nonlinear programming, mathematical programming. Math. Program. **106**(1), 25–57 (2006)

61. Najm, W.G., Smith, J.D., Yanagisawa, M., et al.: Pre-crash scenario typology for crash avoidance research. United States. National Highway Traffic Safety Administration (2007). https://rosap.ntl.bts.gov/view/dot/6281
62. Ferguson, D., Howard, T.M., Likhachev, M.: Motion planning in urban environments. J. Field Robot. **25**(11–12), 939–960 (2008)
63. Afolabi, O., Driggs–Campbell, K., Dong, R., Kochenderfer, M.J., Sastry, S.S.: People as sensors: imputing maps from human actions. In: IROS, pp. 2342–2348, October 2018
64. Naser, F., Gilitschenski, I., Amini, A., Rus, D.: Infrastructure-free NLoS obstacle detection for autonomous cars. In: IROS, pp. 250–257, November 2019
65. McGill, S.G., et al.: Probabilistic risk metrics for navigating occluded intersections. IEEE Robot. Autom. Lett. **4**(4), 4322–4329 (2019)
66. Bühler, A., Gaidon, A., Cramariuc, A., Ambrus, R., Rosman, G., Burgard, W.: Driving through ghosts: behavioral cloning with false positives, August 2020
67. Kochenderfer, M.J.: Decision Making Under Uncertainty: Theory and Application. MIT Press, Cambridge (2015)
68. Yang, X., Liu, L., Vaidya, N.H., Zhao, F.: A vehicle-to-vehicle communication protocol for cooperative collision warning. In: The First Annual International Conference on Mobile and Ubiquitous Systems: Networking and Services, 2004. MOBIQUITOUS 2004, pp. 114–123, August 2004
69. Huang, X., McGill, S.G., Williams, B.C., Fletcher, L., Rosman, G.: Uncertainty-aware driver trajectory prediction at urban intersections. In: 2019 International Conference on Robotics and Automation (ICRA), pp. 9718–9724, May 2019
70. Huang, X., McGill, S.G., DeCastro, J.A., Fletcher, L., Rosman, G.: DiversityGAN: diversity-aware vehicle motion prediction via latent semantic sampling. IEEE Rob. Autom. Lett. **5**(4), 1 (2020)
71. Huang, X., et al.: CARPAL: confidence-aware intent recognition for parallel autonomy. RA-L **6**, 4433–4440 (2021)
72. Huang, X., et al.: HYPER: learned hybrid trajectory prediction via factored inference and adaptive sampling. arXiv [cs.RO], 05 October 2021. http://arxiv.org/abs/2110.02344
73. Huang, X., Rosman, G., Jasour, A., McGill, S.G., Leonard, J.J., Williams, B.C.: TIP: task-informed motion prediction for intelligent systems. arXiv [cs.RO], 17 October 2021. http://arxiv.org/abs/2110.08750
74. Kuo, Y.-L., et al.: Trajectory prediction with linguistic representations, arXiv [cs.RO], 19 October 2021. http://arxiv.org/abs/2110.09741
75. Shalev-Shwartz, S., Shammah, S., Shashua, A.: On a formal model of safe and scalable self-driving cars. arXiv [cs.RO], 21 August 2017. http://arxiv.org/abs/1708.06374
76. Corso, A., Kochenderfer, M.J.: Interpretable safety validation for autonomous vehicles. In: 2020 IEEE 23rd International Conference on Intelligent Transportation Systems (ITSC), pp. 1–6, September 2020
77. Rudenko, A., Palmieri, L., Herman, M., Kitani, K.M., Gavrila, D.M., Arras, K.O.: Human motion trajectory prediction: a survey. IJRR (2019)
78. Lefèvre, S., Vasquez, D., Laugier, C.: A survey on motion prediction and risk assessment for intelligent vehicles. ROBOMECH J. **1**(1), 1–14 (2014). https://doi.org/10.1186/s40648-014-0001-z
79. Luo, K., et al.: Safety-oriented pedestrian motion and scene occupancy forecasting (2021). http://arxiv.org/abs/2101.02385
80. Li, Y., Yu, R., Shahabi, C., Liu, Y.: Diffusion convolutional recurrent neural network: data-driven traffic forecasting. Presented at the ICLR (2018)
81. Li, X., et al.: Vehicle trajectory prediction using generative adversarial network with temporal logic syntax tree features. IEEE Rob. Autom. Lett. **6**(2), 3459–3466 (2021)

82. Li, X., Rosman, G., Gilitschenski, I., Decastro, J.A., Rus, D.: Differentiable logic layer for rule guided trajectory prediction, March 2021
83. Deo, N., Trivedi, M.M.: Multi-modal trajectory prediction of surrounding vehicles with maneuver based LSTMs (2018)
84. Wulfmeier, M., Ondruska, P., Posner, I.: Maximum entropy deep inverse reinforcement learning. arXiv [cs.LG], 17 July 2015. http://arxiv.org/abs/1507.04888
85. Rhinehart, N., Mcallister, R., Kitani, K., Levine, S.: PRECOG: PREdictions conditioned on goals in visual multi-agent scenarios. Presented at the ICCV (2019). Accessed 6 Jan 2022. ICCV
86. Li, L.L., et al.: End-to-end contextual perception and prediction with interaction transformer, August 2020
87. Ettinger, S., et al.: Large scale interactive motion forecasting for autonomous driving: the Waymo Open Motion Dataset. arXiv [cs.CV], 20 April 2021. http://arxiv.org/abs/2104.10133
88. Chang, M.-F., et al.: Argoverse: 3D tracking and forecasting with rich maps. Presented at the CVPR (2019)
89. Houston, J., et al.: One thousand and one hours: self-driving motion prediction dataset. Presented at the CoRL (2020)
90. Liang, M., et al.: Learning lane graph representations for motion forecasting. In: Vedaldi, A., Bischof, H., Brox, T., Frahm, J.-M. (eds.) ECCV 2020. LNCS, vol. 12347, pp. 541–556. Springer, Cham (2020). https://doi.org/10.1007/978-3-030-58536-5_32
91. Gao, J., et al.: VectorNet: encoding HD maps and agent dynamics from vectorized representation. Presented at the CVPR, June 2020
92. Rasouli, A., Kotseruba, I., Tsotsos, J.K.: Agreeing to cross: how drivers and pedestrians communicate. Presented at the IVS (2017)
93. Rasouli, A., Kotseruba, I., Kunic, T., Tsotsos, J.: PIE: a large-scale dataset and models for pedestrian intention estimation and trajectory prediction. Presented at the ICCV (2019)
94. Kotseruba, I., Rasouli, A., Tsotsos, J.K.: Benchmark for evaluating pedestrian action prediction. Presented at the WACV, January 2021
95. Liu, B., et al.: Spatiotemporal relationship reasoning for pedestrian intent prediction. RA-L (2020)
96. Martin, S., Vora, S., Yuen, K., Trivedi, M.M.: Dynamics of driver's gaze: explorations in behavior modeling and maneuver prediction. IEEE Trans. Intell. Veh. **3**(2), 141–150 (2018)
97. Gopinath, D., Rosman, G., Stent, S., Terahata, K., Fletcher, L., Leonard, J.: MAAD: a model and dataset for 'Attended Awareness' in driving. Presented at the ICCV (2021). https://github.com/ToyotaResearchInstitute/att-aware/
98. Cunningham, A.G., Galceran, E., Mehta, D., Ferrer, G., Eustice, R.M., Olson, E.: MPDM: multi-policy decision-making from autonomous driving to social robot navigation. https://apps.dtic.mil/sti/pdfs/AD1100223.pdf
99. Casas, S., Sadat, A., Urtasun, R.: MP3: a unified model to map, perceive, predict and plan. Presented at the IEEE/CVF Conference on Computer Vision and Pattern Recognition, Nashville, TN, USA, June 2021
100. Sadat, A., Casas, S., Ren, M., Wu, X., Dhawan, P., Urtasun, R.: Perceive, predict, and plan: safe motion planning through interpretable semantic representations. In: Vedaldi, A., Bischof, H., Brox, T., Frahm, J.-M. (eds.) ECCV 2020. LNCS, vol. 12368, pp. 414–430. Springer, Cham (2020). https://doi.org/10.1007/978-3-030-58592-1_25
101. Ivanovic, B., Pavone, M.: Rethinking trajectory forecasting evaluation. arXiv [cs.RO], 21 July 2021. http://arxiv.org/abs/2107.10297
102. O'Kelly, M., Sinha, A., Namkoong, H., Duchi, J., Tedrake, R.: Scalable End-to-End autonomous vehicle testing via rare-event simulation. Presented at the NeurIPS (2018)

103. Cao, Z., et al.: Reinforcement learning based control of imitative policies for near-accident driving. Presented at the RSS, July 2020
104. Caesar, H., et al.: nuScenes: a multimodal dataset for autonomous driving. Presented at the CVPR, June 2020
105. Phan-Minh, T., Grigore, E.C., Boulton, F.A., Beijbom, O., Wolff, E.M.: CoverNet: multimodal behavior prediction using trajectory sets. Presented at the CVPR (2020)
106. Shiroshita, S., et al.: Behaviorally diverse traffic simulation via reinforcement learning. In: IROS, pp. 2103–2110, October 2020
107. Gilitschenski, I., Rosman, G., Gupta, A., Karaman, S., Rus, D.: Deep context maps: agent trajectory prediction using location-specific latent maps. IEEE Robot. Autom. Lett. **5**(4), 5097–5104 (2020)
108. Li, X., et al.: Learning an explainable trajectory generator using the automaton generative network (AGN). RA-L **7**(2), 984–991 (2022)
109. Mulder, M., Abbink, D.A.: Sharing control with elderly drivers: haptic guidance during curve negotiation. IFAC Proc. Vol. **43**(13), 310–315 (2010)
110. Campbell, J.L., et al.: Human factors design guidance for level 2 and level 3 automated driving concepts (2018). https://trid.trb.org/view/1574671

Automated Shuttles and Buses for All Users

Katherine F. Turnbull(✉)

Texas A&M Transportation Institute, 3135 TAMU, College Station, TX, USA
k-turnbull@tti.tamu.edu

Abstract. This chapter presents information on demonstrations and pilots of automated shuttles and buses, with a focus on improving transportation options for all users, including individuals with disabilities. Automated shuttles and buses are being piloted, demonstrated, and deployed in downtown areas, university campuses, business parks, entertainment complexes, and other areas. The chapter is based on the presentations and discussions at a breakout session at the 2021 Transportation Research Board (TRB) Automated Road and Transportation Symposium (ARTS). Like many things in life, automated shuttle and bus projects were put on hold or pivoted to food delivery and alternative uses during the pandemic in 2020 and 2021. The ARTS breakout session, Automated Shuttles and Buses for All Users, highlighted addressing the needs of individuals using wheelchairs, those with limited or no eyesight, and those with other disabilities. The information presented in this chapter will assist in evaluation of this mobility option to help inform decision making, identify research needs, and support future developments.

Keywords: Autonomous shuttles · Driverless shuttles · Automated shuttles · Automated buses · Autonomous buses · Driverless buses

1 Introduction

Many pilots, demonstrations, and deployments of automated shuttles and buses in the United States and other countries were put on hold or pivoted to alternative uses during the pandemic. Service resumed on existing projects and new pilots were initiated by mid-2021. These services focus on enhancing mobility and accessibility on regular routes, providing first- and last-mile trips, and improving transportation options for individuals with disabilities. The ARTS 2021 breakout session highlighted pilots and demonstrations in Ohio, Florida, Texas, Minnesota, North Carolina, and Wyoming. A project in Scotland was also discussed. Research projects being undertaken by universities and institutes in Texas, Florida, and Pennsylvania were summarized. Participants in the session shared experiences with automated shuttles and buses, highlighted outreach efforts and use by disabled individuals, and described lessons learned and tips for others interested in similar applications. The involvement of the private sector in enhancing mobility with automated shuttles and buses was also discussed.

G. Meyer and S. Beiker (Eds.): ARTSymposium 2021, LNMOB, pp. 110–119, 2023.
https://doi.org/10.1007/978-3-031-11112-9_9

2 Examples of Automated Shuttle and Bus Pilots, Demonstrations, and Deployments

2.1 Linden Leap, Columbus, Ohio

The pandemic changed the mission of the One Linden project, which was part of the Columbus Smart City program. The initial mission was to use self-driving shuttles to close transportation gaps in reaching public transit, affordable housing, healthy food, childcare, recreation, and education. A 2.8-mile loop connected St. Stephen's Community House, the Douglas Community Center, the Rosewind Resident Council, and the Linden Transit Center. One Linden filled a gap in transit service in the area, providing a first-mile/last-mile link.

The electric EasyMile vehicles accommodated up to 15 passengers and were Americans with Disabilities Act (ADA) accessible. Safety operators were on board the vehicles. The vehicles were wrapped with "One Linden—Our Community, Our Future" to help with marketing and public education. The service began in late 2019.

Service was suspended in February 2020 when one of the shuttles braked for an unknown reason, resulting in a passenger falling from their seat. The National Highway Traffic Safety Administration (NHTSA) ordered EasyMile to suspend operating Linden Leap service, as well as 16 other EasyMile shuttles around the country. The review conducted by EasyMile found that an internal safety mechanism was triggered as the shuttle pulled away from a stop, activating the vehicle's brake. In May 2020, the National Highway Traffic Safety Administration (NHTSA) outlined conditions to lift the suspension, which included implementing a new safety passenger enhancement plan developed by EasyMile and NHTSA.

Given the conditions in May 2020 resulting from the pandemic, Linden Leap pivoted from passenger service to food pantry service when the suspension was lifted. Prepackaged food in boxes and bags was delivered to Franklin County residents along a route from St. Stephen's Community House to the Rosewind Community Center. The food pantry service was operated Monday through Friday from noon to 3:00 p.m. from July 30, 2020, to April 1, 2021. During that time, approximately 30,000 meals and 15,000 masks were delivered to residents.

The two EasyMile shuttles operated more in manual mode during the food delivery service since operators prioritized speed over operating in automated mode. More frequent obstacles, including parked cars and queues of vehicles waiting for the food delivery at parking lots also influenced the use of manual operations. Termination of the Linden Leap food delivery service in April 2021 coincided with the conclusion of the Columbus Smart City Project.

2.2 I-STREET, Gainesville, Florida and the Beep, Lake Nona, Florida

Implementing Solutions from Transportation Research and Evaluating Emerging Technologies (I-STREET) at the University of Florida (UF) provides a real-world connected and automated vehicle (AV) testbed in Gainesville focusing on mobility and safety, data analytics, and human factors.

The Gainesville autonomous shuttle pilot project is a partnership with UF and the Florida Department of Transportation (FDOT). The City of Gainesville is also a partner in the project. FDOT is providing funding for the project. Transdev is the service provider, and EasyMile is the vehicle provider.

The autonomous shuttle route links the UF campus and downtown Gainesville. The project was put on hold in 2020 due to the pandemic and NHTSA's suspension of EasyMile shuttle operations. Service was restored in January 2021.

Before and after surveys were conducted to gain insights into the perception of riders. The before survey was conducted online and in person in the summer of 2018. The after survey was conducted online in the spring of 2021. The surveys included questions on travel behavior and technology use, autonomous shuttle comfort and safety, and demographics. Perceptions related to riding in the shuttles and the ability of the service operator to take control of the shuttle were much more positive in the after surveys once individuals rode in the vehicles. Some concerns over equity, based on the location of the route, were noted in the after surveys.

Florida I-STREET is also evaluating the Beep shuttle operating at Lake Nona, which represents a partnership between Beep and TAVISTOCK Development Company. The service, using a NAVYA shuttle, launched in September 2019. The service was halted due to the pandemic in March 2020. Service was relaunched in June 2020.

The evaluation of the Lake Nona shuttle includes collecting and analyzing vehicle trajectories and design characteristics to assess the action of the AV shuttle while interacting with surrounding traffic. Three intersections with different characteristics were examined through in vehicle videos and videos of the intersection operations with and without the shuttle.

Both projects highlight the importance of stakeholder involvement and interaction. Regular communication was used on both projects to ensure all groups were actively engaged in all aspects of the pilots. Providing current information on the pilots was a key element of the communication process.

2.3 Arlington Rideshare, Automation, and Payment Integration Demonstration (RAPID), Arlington, Texas

The City of Arlington has conducted two AV shuttle pilots and has a third underway. Milo, which operated on off-street trails in the Arlington Entertainment District from August 2017 to August 2018, was the first autonomous shuttle in the country offered by a municipal government to the public on a continuous basis. A trained operator was always on board and could take control of the vehicle if needed. Rides were free. The two major goals of Milo were to test the AV technology in a real-world environment and to educate the public on AVs. Milo served over 110 events, including 78 stadium events, 17 public demonstrations, and 18 special group tours.

The second pilot was on-street operation of Drive.ai AVs, which operated from October 2018 to May 2019. It included Drive.ai vehicles operating in mixed traffic at speeds up to 35 mph. Rides were free, and the service was open to the public. A total of 755 AV trips were made, serving 1,424 passengers. A total of 451 AV passenger miles were driven.

The current pilot resulted when the city was awarded $1.7 million from the Federal Transit Administration Integrated Mobility Innovation competitive grant program for Arlington Rideshare, Automation, and Payment Integration Demonstration (RAPID). Partners on the project include Via, May Mobility, and The University of Texas at Arlington (UTA). The project integrates May Mobility AVs with Via's on-demand rideshare platform in the downtown area and on the UTA campus. The one-year demonstration began in March 2021.

The AV fleet includes four Lexus hybrid electric sedans and one electric Polaris GEM that is wheelchair accessible. An attendant is always behind the wheel and transitions between manual and automated mode. Additional health safety features have been implemented, including a partition between the attendant and the passenger areas and daily cleaning. Riders are currently required to wear masks.

The one square-mile RAPID service area encompasses the downtown and the UTA campus. Service is provided Monday through Friday from 7:00 a.m. to 7:00 p.m. Trip booking is accomplished using the Via app or reservation telephone number. Riders have a choice of requesting the RAPID AVs or the standard Via vans. The standard Via fares apply—$3–$5 per person per ride or $25 for a weekly pass good for up to four rides a day. UTA students ride for free.

There has been a steady increase in ridership since the start of service. The average daily ridership grew from 37 riders in March to 65 in June. In addition, the integration with Via for booking, payment, vehicle dispatch, and routing has been successful. The automated performance of the vehicles has been good throughout the service area.

2.4 The Med City Mover, Rochester, Minnesota

The Med City Mover was selected through the Minnesota Department of Transportation (MnDOT) CAV Challenge Program. It includes the operation of two EasyMile EZ10 vehicles in downtown Rochester, which is the home of the Mayo Clinic. The demonstration, which was delayed due to the pandemic, began service in September 2021. The service is open to the public and will be in operation for 12 months.

MnDOT is the project lead. The project partners include the City of Rochester, the Mayo Clinic, and Destination Medical Center. First Transit and EasyMile are the technology and transit operations partners.

The project has four goals focusing on public education, winter weather, infrastructure, and mobility. The first goal is to engage Minnesotans about the potential benefits and opportunities of AV technology. The second goal is to improve the operation of AVs in winter weather conditions. The third goal is to identify changes to infrastructure needed to safely operate AVs on public roadways. The fourth goal is to enhance the transportation experience for Rochester residents, businesses, and visitors, and improve how people get around in the high-demand downtown area.

Two six-passenger electric EasyMile E210 shuttles operate on a 1.5-mile-long loop, in the central part of the downtown area. The route connects the May Clinic Downtown Campus with restaurants, grocery stores, residential areas, apartment complexes, hotels, and parking facilities. The route includes stops at two locations. The free service is provided from 9:00 a.m. to 3:00 p.m., Monday–Friday, and 9:00 a.m. to 5:00 p.m. Saturday and Sunday.

There is an onboard ambassador to assist passengers. The ambassador can take over operation of the shuttle if needed. The EasyMile EZ10 vehicles and the project include a number of accessibility features. The vehicles have wheelchair ramps and wheelchair tie-downs. The shuttles also have signage in Braille, and use audio messages, trolly bells, and video with closed captioning to communicate with disabled passengers. The Med City Mover will operate for a 12-month period. MnDOT and project partners will evaluate the operation of the service, vehicle performance, rider feedback, and other elements.

2.5 Automated Shuttles in the Wright Brothers National Memorial and Yellowstone National Park

During the summer of 2021, the National Park Service (NPS) piloted automated shuttles at the Wright Brothers National Memorial in North Carolina and Yellowstone National Park in Wyoming. The NPS is dedicated to conserving unimpaired the natural and cultural resources and values of the National Park System for the enjoyment, education, and inspiration of this and future generations. Transportation to, from, and within the National Park System is key to achieving this vision.

The NPS has implemented innovative transit services in many parks over the past 20 years. Examples include bus systems in Zion and Acadia National Parks, micro mobility on the National Mall, ride hailing in Golden Gate and Great Falls National Parks, and real-time traveler information in Acadia and Bryce Canyon National Parks. The NPS created an emerging Mobility Working Group in late 2019.

The automated shuttles at the Wright Brothers National Memorial and Yellowstone National Park represents the most recent innovative transportation services piloted by the NPS. The NPS AV shuttle pilots serve several purposes. First, they demonstrate the use of AV shuttle technologies in novel operating environments, including rural and remote areas and recreational settings. Second, the projects help identify and overcome unforeseen regulatory, organizational, and legal barriers to further deployments of AV shuttles in the unique settings. Third, they enhance visitor experience by facilitating new interpretive opportunities and improving mobility assistance. It is also important to note that the projects are not intended to address traffic congestion or test transit sustainability for high-capacity service.

The Connected and Autonomous Shuttle Supporting Innovation (CASSI) pilot operated at the Wright Brothers National Memorial in North Carolina from April 20 to July 12, 2021. Two EasyMile shuttles were operated by TransDev. The pilot was a joint partnership between NPS and the North Carolina Department of Transportation (NCDOT). The CASSI service operated on a 1.5-mile loop around the Wright Brothers Memorial from the museum stop to the sculpture stop. Service was provided daily from 10:00 AM to 4:30 PM.

The TEDDY electric driverless shuttle demonstration operated in Wyoming's Yellowstone National Park from June 9 to August 31, 2021. Service was provided on two routes in the Canyon Village area over the 84 days. Route 1A operated from June 9 to June 12, providing service from the Canyon Lodge and Cabins, to Canyon Village. Route 1B provided service from the Canyon Campground to Canyon Village from July 14 to

August 31, 2021. The two routes provided the opportunity to serve different markets of park visitors. BEEP provided the service operating Local Motors Olli Shuttles.

The NPS Accessibility Office was involved in the pilots, and both included a number of accessibility considerations. The vehicles did have ramps, allowing access for wheelchair users. Infrastructure enhancements were also made to improve access by wheelchair users. The onboard attendant was available to anyone needing assistance. A video with closed caption message was available for the TEDDY shuttles. Evaluations are being conducted by the NPS and the U.S. Department of Transportation's Volpe National Transportation Systems Center.

2.6 Houston METRO AV Proving Ground, Houston, Texas

The Metropolitan Transit Authority of Harris County (Houston METRO) was the lead on the University District Phase 1 pilot, which included operating an automated shuttle on a 1-mile closed-loop route on the Texas Southern University (TSU) campus in Houston, Texas. The pilot began in June 2019 and ended in February 2020. Project members included the Houston-Galveston Area Council (H-GAC) and TSU. The Texas Department of Transportation, the Texas Innovation Alliance, and the University of Houston also participated in the pilot.

First Transit, Inc. was the shuttle operator, and EasyMile was the shuttle vehicle provider. An attendant was on board to take control if needed. The electric EasyMile 10 Gen 2 vehicle provided capacity for 12 passengers (6 seated and 6 standing) and provided access for passengers with reduced mobility. The Center for Transportation Training and Research at TSU conducted research related to vehicle operations, passenger engagement, workforce needs, environmental impacts, and safety.

Some accessibility challenges were encountered during Phase I. The vehicle could accommodate wheelchair users, but it was not fully ADA compliant. The slope of the wheelchair ramp was challenging, especially for heavy wheelchairs. The entry step was high, which created issues for some individuals. The lack of curb cuts and other concerns with the built environment also caused problems for some individuals with limited mobility.

METRO was selected for the FTA's Accelerating Innovative Mobility (AIM) Challenge Grant to implement Phase II, which will connect to the Purple Light Rail Transit (LRT) station, serving TSU, the University of Houston, and the Third Ward community. METRO's Phase II will use a Phoenix Motorcar Zeus 400 Motor F-450 Chassis, which is a mid-size vehicle. It is ADA, NHTSA, and Buy America compliant. In addition to Phoenix Motorcars, other project partners include EasyMile and AECOM.

2.7 CAVForth, Edinburgh Scotland

This project is developing a fleet of five automated buses and operating them in high-capacity service on a 14-mile route across the Forth Road Bridge that links the Ferrytoll park-and-ride facility with the Transport Hub at Edinburgh Park Station. CAVForth project partners include Fusion Processing, Stagecoach, Bristol Robotics Laboratory, and Edinburgh Napier University.

While the project was put on hold during the pandemic, a survey was conducted to gauge interest in public transit and automated buses after the pandemic. Many people reported that they were not well informed about automated buses. Further, they indicated that automated buses would not encourage them to use public transit more often. Some respondents reported that they would wait until more experience was gained with automated buses, rather than being an early user. Respondents responded that they would be more likely to use an automated bus with the presence of a driver and a steward than options with no driver and a steward and no driver and no steward.

3 Research Projects

3.1 Automated Shuttles and Buses for All Users, Texas A&M Transportation Institute

The research project, Automated Shuttles and Buses for All Users, is being conducted by the Texas A&M Transportation Institute (TTI) as part of the Safety Through Disruption (SAFE-D) University Transportation Center (UTC) led by the Virginia Tech Transportation Institute. The project is introducing individuals with mobility and visual impairments to AVs and a smart intersection and gaining information on their complete trip. The project is identifying improvements in AVs, service operations, the street system, and the built environment to ensure that individuals with disabilities have equal and safe access to automated shuttles and buses to improve their mobility.

In cooperation with the City of Arlington, UTA, Via Rideshare, and May Mobility, the project is introducing individuals with mobility and visual impairments to the RAPID automated shuttles. An initial session was conducted on June 17, 2021, with five individuals. The session included a pre-interview, a ride in a RAPID shuttle, and a post-interview.

The participants and the RAPID vehicle used are highlighted below:

- Young male using a wheelchair with a companion—GEM van
- Middle-aged female with no sight with a guide dog—Lexus sedan
- Middle-aged male with colorblindness—Lexus sedan
- Older female using a walker with a companion—Lexus sedan
- Older male using a walker—Lexus sedan

The overall initial reactions to riding in the RAPID shuttles from the individuals was very positive. All five noted they felt comfortable and safe riding in the vehicles. They noted the generally smooth ride, including when the vehicle was in automated operation. Three of the five participants used assistance to enter and exit the vehicles, and they noted the importance of having an assistance for regular use. All five participants noted they would use the service on a regular basis. The blind individual stressed the need to ensure and verify pickup and drop-off locations. It was also suggested that similar vehicles for wheelchairs users would ensure equity in the services.

Participants also provided insights into their complete trips and the built environment. They noted the importance of handicapped vehicle parking spots, as well as curb cuts, ramps, and accessible sidewalks. The need to lower traffic speeds in many areas was noted. Ensuring the correct locations for pickups and drop-offs and the need for on-vehicle attendants to provide assistance was stressed as important. Questions and concerns on sharing vehicles with others were noted by some participants.

Future activities on the project include conducting additional rides using the RAPID shuttles in the fall with UTA students, faculty, and staff. Focus groups on the smart intersection will be conducted in the fall in College Station. The results of these activities will be used to identify enhancements to aid disabled individuals' access to and use of AVs. The project will host workshops and develop videos to highlight the results.

3.2 Demonstration Study: Older Drivers' Experiences with Autonomous Vehicle Technology, University of Florida

The Older Drivers' Experiences with Autonomous Vehicle Technology study was conducted by the Institute for Mobility, Activity and Participation at the University of Florida. The project was funded through the Southeastern Transportation Research, Innovation, and Education (STRIDE) UTC. Project stakeholders included Transdev EasyMile, the City of Gainesville, Oak Hammock Residential Community, Rotary Clubs Gainesville, UF Transportation Institute, FDOT, and FLSafe Mobility for Life Coalition.

The objectives of the project were to develop and validate an Automated Vehicle User Perception Survey (AVUPS) and to obtain and analyze information on the perceptions of older drivers about AV technology before and after "driving" a simulator and riding in an automated shuttle.

Participants in the project were required to be 65 years of age or older and hold a valid driver's license. A total of 104 participants were included in the project. The project equipment included a RTI high-fidelity driving simulator and a EasyMile EZ10 automated shuttle.

To document older drivers' perceptions toward AVs, the study used a repeated measures crossover design, with random allocation of 104 older drivers who were exposed to (a) an autonomous shuttle (Society of Automotive Engineers Level 4) and (b) a simulator programmed to run in autonomous mode (Society of Automotive Engineers Level 4). Participants completed pre- and post-exposure surveys, to report their adoption preferences and perceptions on nine domains including the AVUPS.

A two-way mixed analysis of variance was used to analyze the time effect, group effect, and time by group interaction. No group effects were evident, but older drivers' perceptions of safety, trust, and perceived usefulness of AV technology increased after being exposed to the AV technology. Examination of the group by time interaction effects indicated the significance of older adult perceptions pertaining to intention to use, trust, perceived usefulness, control/driving efficacy, and safety.

The study provides valuable contributions to the current body of knowledge regarding the determinants of older adult AV technology acceptance practices. It also highlights that repeated testing would be beneficial to better understand how different automated systems, levels of technology, contexts, policies, and local conditions may influence older drivers' perceptions of AV technology, including automated shuttles.

3.3 Automated Vehicle Services for People with Disabilities—Involving Responsive Engineering Center, University of Pittsburgh

The Automated Vehicle Services for People with Disabilities-Involving Responsive Engineering (ASPIRE) Center is a UTC funded by the U.S. Department of Transportation based at the University of Pittsburgh's Human Engineering Research Laboratories. Other members of the ASPIRE Center include the Catholic University of America and the Uniformed Services University of Health Sciences.

The ASPIRE Center project is focused on developing a road map for use by service providers and manufacturers that addresses the needs of disabled individuals and assists with integrating accessible AVs and mobility services. The three major project activities are conducting a systematic literature review, gathering input from consumers and providers, and modeling use levels by people with disabilities.

The literature review identified 27 scientific articles and reports, and 34 agency reports and other documents. The term "grey" literature is often used for this second group of reports since they are not always recorded in databases maintained by the TRB and other organizations. Literature focusing on travel of older adults was the most prevalent. Some of the findings from the literature review included the need to focus on the entire travel journal and tailoring solutions to the needs and preferences of individuals with different disabilities and impairments.

The findings also highlight the need to develop guidance on accessibility and design for use in planning AV technologies and supporting infrastructure. The literature review was used to identify research gaps and implications for policy and knowledge translation. The research gaps identified included the lack of information on transportation trends and socio-demographic factors, accessibility and useability of AVs and related services by disabled individuals, and the outcomes of AV use by the disabled community. The need for additional research on these and other topics was supported by the literature review results.

One of the policy implications identified in the literature review was the need to include universal design and participatory action design and engineering principles as part of the development of AVs, AV services, and the built environment. A second policy implication focused on the payment for services, state and federal coverage, and voucher systems, especially given the impact AVs may have on access to health care, employment, and education. Developing policies to address the driver licensing requirements for AVs to ensure disabled individuals are not excluded is also important.

Focus groups and journey mapping, a pilot survey, and a nationwide survey are being used to gain insights from disabled consumers as well as service providers. The journey mapping questions focus on obtaining information on challenges or difficulties experienced by disabled individuals during each step of the trip. Information on what assistance is needed and what challenges or improvements would make that phase easier is being gathered. Questions related to the impact of different origins and destinations on the travel process and how disabled individuals adjust to problems encountered during a trip are also included.

The focus groups will also obtain general information on how AVs might eliminate transportation issues for disabled individuals, perceptions on barriers to using AVs, and any safety concerns with AV use. Focus group participants will also be asked about possible benefits from AVs.

4 Private Sector Activities with Automated Shuttles and Buses

Although different business models are being used to plan and operate automated shuttles and buses, the private sector is actively involved in all projects. Automated vehicle companies, transit service operators, land development companies, and major businesses provide examples of the diverse private sector interest in enhancing mobility through automated shuttles and buses.

Electric automated shuttles developed by EasyMile, Local Motors, NAVA, Drive.ai, Lexus hybrid sedans, and Polaris GEM vans have been used in various pilots and demonstrations. First Transit, TransDev, and BEEP provide examples of transit service companies operating the services. Land and community development companies, major employers, and businesses have also partnered on demonstration projects. These groups bring technical and operating expertise, resources, and business strengths to the various projects.

5 Additional Research

A number of areas for further research projects, pilots, and evaluation have been highlighted through projects, studies, and discussions at the AVS breakout sessions. Participants discussed the need for further research focusing on the use of automated shuttles and buses by disabled individuals. Outreach to the disabled community was highlighted, along with conducting more pilots and demonstrations addressing the needs of disabled individuals.

Other examples of topics for additional research include on-road and on-vehicle signing, sensor and battery robustness and performance, and remote supervision and monitoring. Additional topics include common evaluation methodologies and core questions for user and public surveys. Continuing to share experiences with pilots, demonstrations, and deployments will also be important.

Part IV: Vehicle and Road Systems Technology Development

What's Next in AV Standards and Simulation Validation?

Shawn Kimmel[1](✉), George Nicols[2], Edward Straub[3], and Edward Chow[4]

[1] Quantitative Scientific Solutions, 208 E Street NE, Washington, DC 20002, USA
shawn.c.kimmel@gmail.com

[2] Toyota Motor North America, 1555 Woodridge Avenue, Ann Arbor, MI 48105-9748, USA
george.nicols@toyota.com

[3] SAE ITC, 755 W. Big Beaver Road, Troy, MI 48084, USA
Edward.Straub@sae-itc.org

[4] NASA Jet Propulsion Laboratory (JPL), 4800 Oak Grove Drive, Pasadena, CA 91109, USA
Edward.T.Chow@jpl.nasa.gov

Abstract. Standards have the potential to both enable and hinder the deployment of AVs. With an increasing number of standards acronyms and numbers, it can be challenging for practitioners to make sense of past, present, and future AV standards. The chapter is organized by four "hot topics", including (1) *safety assurance*, (2) *physical infrastructure*, (3) *connectivity and cooperative driving automation,* and (4) *simulation validity and representativeness*. Perspectives from leaders in the standards space are provided, giving the latest on standards activities and insights into where the field is going. Public sector and industry participants, including from US DOT, NIST, NASA, Euro NCAP, Toyota, Advocates for Highway and Auto Safety, Qualcomm Technologies, Inc., Foretellix, Japan Automobile Research Institute, Aurora, Intel / Mobileye, dSPACE Inc, and BMW, described the relevance of standards to their organizations and job function. A globally representative sample of standards development organizations (SDOs), including from SAE, ISO, IEEE, ASAM, and CSA provided their perspectives on how to harmonize the myriad of activities on related and overlapping topics. Standards gaps and future priorities were identified through collaborative, interactive breakout sessions.

Keywords: Automated vehicle · Automated driving system · Standards · Safety assurance · Validation and verification · Physical infrastructure · Connected vehicle · Cooperative driving automation · Simulation · Representativeness · MUTCD · Scenario testing

1 Introduction

These sessions sought to make sense of the myriad of standards activities, which are rapidly increasing in quantity and pace of development, and how standards can enable AV testing, simulation and deployment. The technology for AVs is still evolving, as is

G. Meyer and S. Beiker (Eds.): ARTSymposium 2021, LNMOB, pp. 123–139, 2023.
https://doi.org/10.1007/978-3-031-11112-9_10

the lexicon to describe and ontologies to organize AV concepts. Tracking and understanding standards activities and gaps has become more challenging, time consuming, and confusing.

The proliferation in standards is indicative of the increased value organizations place on standards. Standards have become a key part of market strategies for AV technology developers and system integrators, and provide a foundation for interoperable testing and simulation. The public sector finds value in the way standards can promote trust and scalability of the technology. The sessions were organized by representatives from a globally representative sample of Standards Development Organizations (SDOs), including SAE, ISO, IEEE, ASAM, and public and private sector organizations, including US Department of Transportation, Toyota, Advocates for Highway and Auto Safety, Qualcomm Technologies, Inc., Foretellix, Japan Automobile Research Institute, Aurora, Intel / Mobileye, and BMW. Participants in the breakout session represented those who were interested in how standards can help accelerate deployment of new technologies, including improving performance, facilitating testing and simulation, and capturing other benefits.

Breakout sessions on these topics have been conducted for the past several years, and this year we took a fresh look at the hottest topics, including areas of controversy. The idea for this session was born out of discussions with AV stakeholders that expressed exasperation with trying to understand what activities they needed to pay attention to and integrating standards into their organizational, technical and business planning and strategy. In discussions amongst session organizers, four topics in particular stood out as particularly active: *safety assurance*, *physical infrastructure*, *connectivity and cooperative driving automation,* and *simulation validity*.

Within these topic areas, the sessions sought to make sense of the proliferation of standards activities and what was needed to address clear and present challenges. Speakers explored areas where multiple activities appear to address related or overlapping issues, and how evolving nomenclature creates confusion about how concepts may align or conflict. We clarify the different types of standards products (e.g., best practice, recommended practice, technical standard, regulatory standard), and their role in controlling the deployment of the technology. We discussed standards that are on the immediate horizon, as well as those that are further in the future but will require planning, coordination and considerable energy. The variety of public and private sector speakers described how standards impact their organizations differently. With a variety of AV technologies and business models, this session highlighted the influence of the operational design domain (ODD) on the different standards and regulations.

These concepts were addressed through presentations from experts, and then through moderated discussions with participants. Session moderators reinforced the objectives throughout discussions, and captured key findings that are provided in the sections below.

1.1 Standards Landscape Overview

Dr. Shawn Kimmel from Quantitative Scientific Solutions (QS-2), Vice-Chair of SAE On-Road Automated Driving Committee, provided an overview of the standards landscape using SAE's collaborative standards roadmap, CAV Source, shown in Fig. 1.

This standards tracking and planning tool covers the following topic areas: accessibility, cooperative-ADS, cybersecurity, data, human factors, infrastructure, and safety. He used this roadmap to show several of the current and potential future standards activities related to the four hot topic areas. For safety assurance, he discussed standards related to behavioral, developmental, functional, and operational safety, including metrics, safety cases. For physical infrastructure, he discussed standards relevant to infrastructure design, assessment / readiness, and maintenance, including particular infrastructure elements, such as markings, signage, curb use, work zones, and signalized control. For connectivity and cooperative driving automation, key standards areas include definition of terms, architecture, and use cases. For testing and simulation, key standards activities include scenario-based testing methods, testing operational concepts, middleware, and validation methods.

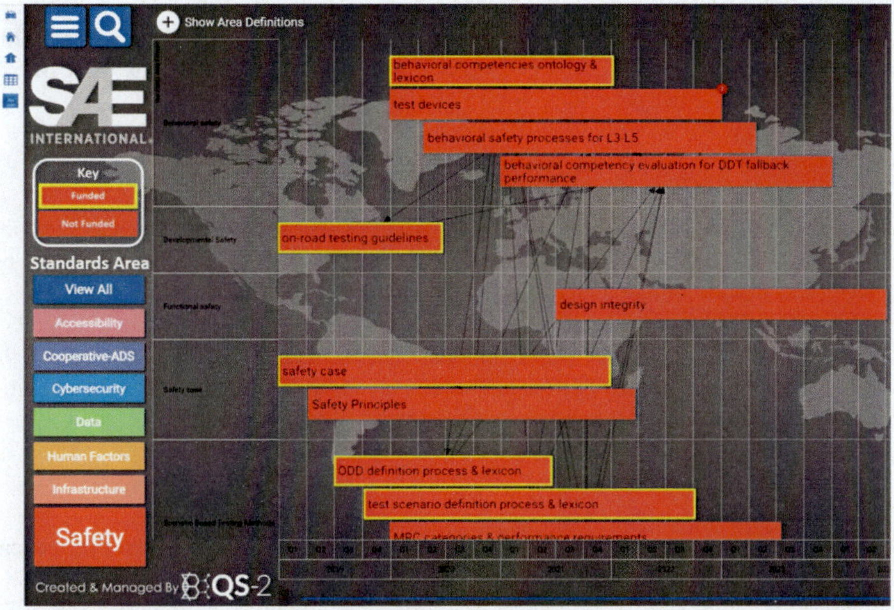

Fig. 1. Standards roadmap showing activities and notional timelines for the development of connected and automated vehcile standards (SAE 2021a).

2 Hot Topic Areas

2.1 Hot Topic 1: ADS Safety Assurance

These discussions explored the latest developments in safety assurance demonstration standards, including methods, notations and scenario-based testing implementation. The session included a series of presentations regarding safety assurance, which set the stage for a corresponding interactive discussion among some of the breakout participants.

Highlights of that discussion were:

1. How can safety requirements be established?

Multiple aspects will likely need to be considered in establishing appropriate requirements. Some topics to consider include creating a basic catalog of driving scenarios, evaluating performance through simulation in addition to physical testing, conducting audits of performance while vehicles are in service, implementing a safety management system, and monitoring vehicle operation.

2. How robust does simulation need to be?

If used for relatively simple cases, maybe simulation does not need to be very robust. This would enable simulations to be employed in an efficient manner. In cases that are more sensitive to parameter values, a high level of robustness may be necessary. This would provide additional confidence in the results obtained. Limitations exist for all types of testing, whether performed by simulation or not.

3. Can simulation-based testing be standardized?

Standardization could be applied to simulation processes rather than what specific simulations are performed. Performing a set of standard specific simulations would not be appropriate because developers could design systems to meet specific tests. A set of specific simulations would be relatively limited in number. However, the total number of possible simulations is practically limitless. Performing specific simulations might not accurately reflect overall performance of a given system. Focusing on simulation processes could allow better characterization of performance.

4. What topics are anticipated to be addressed in ISO TS 5083?

ISO TS 5083 (Road vehicles—Safety for automated driving systems—Design, verification and validation) is intended to address automation at a system level perspective. Other standards documents will be referenced for further details on specific topics. TS 5083 will likely include the concepts of positive risk balance and avoidance of unreasonable risk as methods that could be applied.

This session identified and explored some of the key areas of standards activities, summarized in Fig. 2.

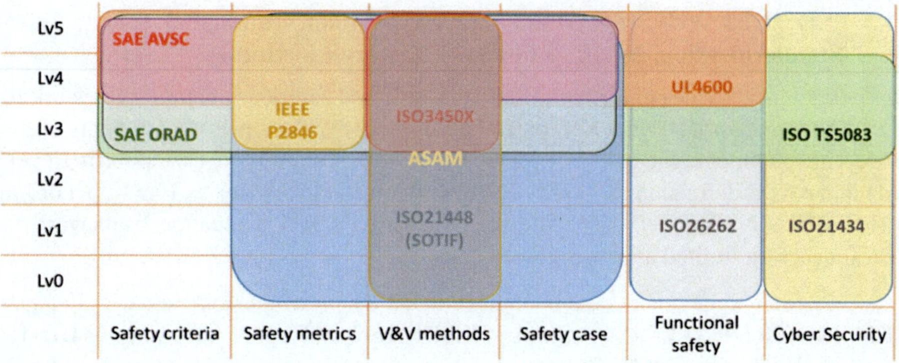

Fig. 2. Safety standards areas and activities (Figure courtesy of Jacobo Antona-Makoshi, JARI, SAKURA project).

2.1.1 ASAM - Ben Engel, Global Technology Manager, ASAM

ASAM is working on a suite of standards to enable scenario-based testing. Ben Engel described several of these standards and their inter-relationships using Fig. 3. He described how OpenODD, OpenSCENARIO (and 2.0), OpenMaterial, OSI, OpenLABEL and OpenXOntology are supporting validation processes, including specification, running tests, postprocessing, storage, and retrieval.

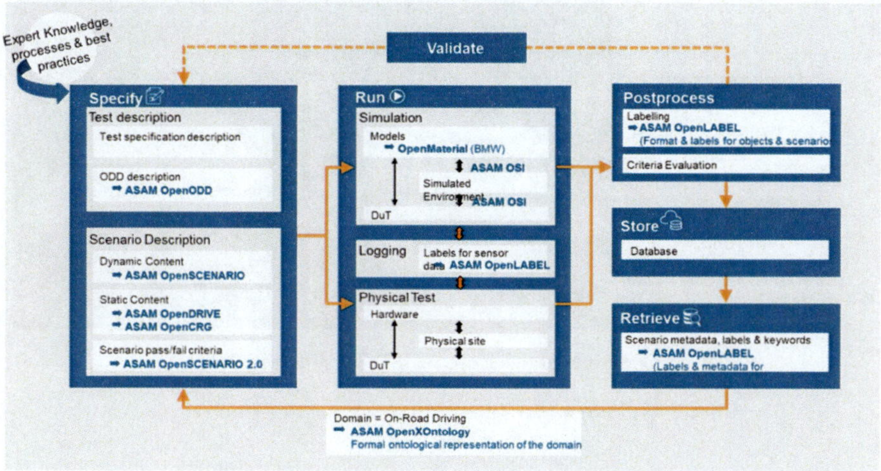

Fig. 3. ASAM suite of standards for scenario-based testing (ASAM 2021).

2.1.2 ISO TC22 WG9, ISO 34502 – Jacobo Antona, Automated Vehicle Safety Standardization, Japan Automobile Research Institute

This document provides guidance for a scenario-based safety evaluation framework for ADS. The framework clarifies a scenario-based safety evaluation process to apply during product development. The guidance for the framework is intended to be applied to Level 3 and higher ADS defined in ISO/SAE 22736 and to vehicle categories 1 and 2 according to (ECE/TR ANS/WP.29/1045). The scenario-based safety evaluation framework for ADS focuses on limited access highways.

2.1.3 ISO TR 5083 / TS 4804 - Simon Fürst, Principal Expert Autonomous Driving Technologies, BMW Group

This document describes steps for developing and validating automated driving systems based on basic safety principles derived from worldwide applicable publications. It considers safety-by-design, as well as verification and validation methods for automated driving systems focused on vehicles with level 3 and level 4 features according to SAE J3016:2018 (see Fig. 4). In addition, it outlines cybersecurity considerations intersecting with objectives for safety of automated driving systems.

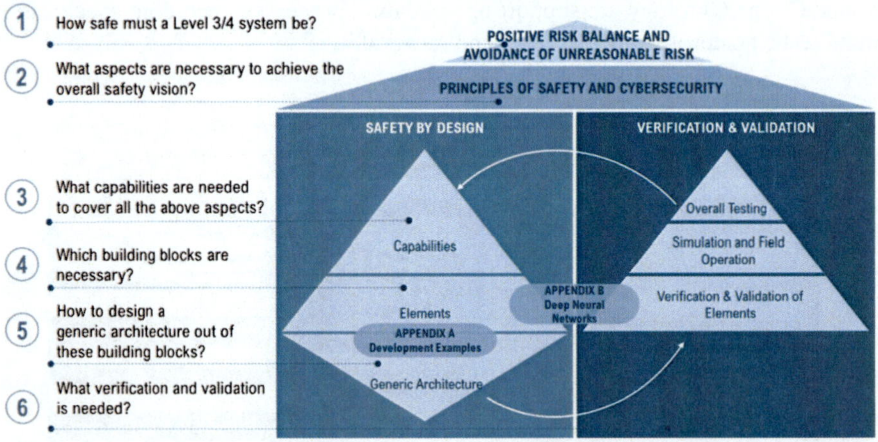

Fig. 4. Overview of the scope of ISO TR 5083 (ISO 2021).

2.1.4 IEEE P2846, SAE ORAD V&V TF - Jack Weast, Chair P2846, Co-chair ORAD V&V TF

IEEE P2846 applies to road vehicles. Within its Normative portion, it defines a minimum set of reasonable assumptions and foreseeable scenarios that shall be considered in the development of safety-related models that are part of automated driving systems (ADS). Jack Weast gave an update on the state of this document and the types of parameters being considered.

2.1.5 UNECE GRVA – Daniel Smith, Assistant General Counsel, Regulatory, Waymo

UNECE is developing international regulations for ADS safety. UNECE includes participants from many nations in Europe, Asia and North America, along with NGOs (industry groups, SAE, etc.). The target is approval of rules in mid-2022 by UNECE, which would later become applicable in member states. The work is organized into several working groups; updates were given on the following working groups:

- FRAV (Functional Requirements for Automated Vehicles) - Starting with high level principles and working toward specific, verifiable requirements.
- VMAD (Validation Methods for Automated Driving) – Organized into subgroups on scenarios, virtual testing, audit and in-use monitoring, and on-road and track testing; each will determine best application to new ADS requirements.
- DSSAD (Data Storage System for Automated Driving) – Developing requirements for data to collect.

Industry standards (including SAE J3016, ISO, ASAM, IEE, etc.) are part of the dialogue but which of those may have direct impact on rules is not yet clear.

2.2 Hot Topic 2: Physical Infrastructure

Infrastructure plays a critical role in safe and efficient traffic operation and will be even more important as automation assumes greater control over the dynamic driving task. The breakout session included presentations of activities from standards development organizations and updates on federal activities. The need for standardized digital communications and visual cues (such as signage and lane markings) was highlighted. The CAVSource tool was used to present several recent and planned physical infrastructure standards (see Fig. 5).

2.2.1 FHWA Activities – John Harding, Connected/Automated Vehicles and Emerging Technologies Team Leader, FHWA

John Harding provided an update on recent efforts relating to physical infrastructure standards related to AVs, including:

- National Roadway Automation Concept of Operations
- Collaborative Framework for Automated Driving Systems/Roadway Testing
- Automated Driving Systems (ADS) Operational Behavior and Traffic Regulation Information
- Cooperative Automated Transportation Coalition (CAT-C)
- Digital Infrastructure Framework (internal)
- International Transport Forum – WG - Road Investments for AV Integration
- Development of Innovative Techniques and Methods that Support a Changeable Roadway Testing Environment

2.2.2 MUTCD Part 5c "What's Next?", SAE ORAD Infrastructure TF – Rob Dingess, President, Mercer Strategic

This presentation provided an overview of upcoming revisions to the Manual on Uniform Traffic Control Devices (MUTCD). The guidance includes a new chapter (Chapter 5) that addresses infrastructure related elements important to automated driving such as digital infrastructure for traffic control devices and recommended standards for road markings such as line widths, gore markings, and contrast. The presentation also reported that FHWA recommends the use of digital Vehicle-to-Everything (V2X) communications standards including recommendations for traffic signal phase and timing (SPaT), work zone data, and rail crossing signal data. The Infrastructure Needs for ADS task force in SAE International's On-Road Automated Driving (ORAD) committee is recommending the use of standards such as the V2X Communication Message SET Dictionary J2735 (2020).

2.2.3 SAE Activities: Edge Report on Open Infrastructure Challenges – Edward Straub, Vice President, Land-Based Systems, SAE ITC

SAE International, a global standards development organization active in automotive and aerospace standards for over 100 years, presented an overview of technical committee work relevant to the infrastructure that can support or facilitate the safe operation of automated driving systems (ADS) and new mobility technologies. This included work in SAE's Cooperative Driving Automation committee, the V2X Vehicular Application & Technical Committee, and committees working on Shared & Digital Micromobility topics. Future work at SAE International includes publishing of a series of infrastructure-related "Edge Reports" (SAE 2021b) which explore unsettled topics in mobility technology. Planned titles will address Unsettled Topics in AVs and Infrastructure: Freight; Future-Proofing Infrastructure Investment; Infrastructure Enablers; and Transit, Infrastructure, and AVs.

2.2.4 CAV Physical and Digital Infrastructure Code – Mahmood Nesheli, Project Manager, CSA Group

The Canadian Standards Association, (CSA Group), is a unique organization, offering over 100 years of experience in the facilitation, project management, and development of consensus-based codes and standards for North America. CSA Group developed a Connected and Automated Vehicle (CAV) Standardization Roadmap that provides a view of the existing standards landscape for topics such as digital and physical infrastructures, data management, privacy, and cybersecurity. It highlights critical requirements and gaps in CAV codes and standards that need to be addressed. CSA also published a report, "Connected and Automated Vehicle Technologies – Insights for Codes and Standards in Canada," which provides detailed insight into the existing standards landscape, the gaps, and the themes that require attention and action (CSA 2020). "Physical and Digital Infrastructure for Connected and Automated Vehicles (CAV)," was also published to further examine the proper framework to integrate CAV digital and physical infrastructure requirements (CSA 2021). CSA Group is also developing a bi-national

CAV Code that will provide a comprehensive contextual framework for CAV deployment and operations by leveraging existing applicable standards. The CAV Code will focus on physical infrastructure and digital infrastructure integration of various systems and provide direction for safety requirements for CAV infrastructure.

2.3 Hot Topic 3: Connectivity and Cooperative Driving Automation

Connectivity and cooperative driving automation (CDA) aim to improve the safety and flow of traffic and/or facilitate road operations by supporting the movement of multiple vehicles and road users. Improved performance is dependent on sharing information that can be used to influence (directly or indirectly) the dynamic driving task (DDT) performance. In order to realize improvements for safety, mobility, situational awareness, and operations, standards are needed to ensure interoperability and trust. This session explored ongoing and planned connectivity and CDA standards activities.

2.3.1 SAE CDA Committee, SAE J3216 – Barb Wendling, Senior Engineer, QS-2, Chair SAE CDA Committee

Barb Wendling described SAE standards efforts related to CDA, defined as machine-to-machine (M2M) communication to enable cooperation between two or more participating entities or communication devices possessed or controlled by those entities. The cooperation supports or enables performance of the dynamic driving task (DDT) for a subject vehicle with driving automation feature(s) engaged. Other participants may include other vehicles with driving automation feature(s) engaged, shared road users (e.g., drivers of manually operated vehicles or pedestrians or cyclists carrying personal devices), or road operators (e.g., those who maintain or operate traffic signals or work zones). Participants can include vehicles, infrastructure elements, and other road users and a variety of information types, including state (e.g., position, signal phase), intent (e.g., planned trajectory), or seek agreement on a plan (e.g., coordinated merge).

2.3.2 V2X and Connected Intersections – Justin McNew, President, JMC Rota, Vice-Chair SAE V2X Communications Steering Committee

Justin McNew described the connectivity and CDA standards landscape (see Fig. 5). There are over twenty-five SAE documents (so far) defining applications that use V2X communications, including vehicle safety and cooperative driving automation. 3GPP PC5 Mode 4 (C-V2X) and 802.11 (DSRC) are the currently supported physical interfaces in IEEE 1609. SAE Standards include J2735 (Data Dictionary), J2945/1 and J3161/1 (Vehicle Safety Communications), etc.

The connected intersections project is sponsored by USDOT ITS Joint Programs Office (JPO), and supported by Standards Development Organizations (SDOs) including Institute of Transportation Engineers (ITE) (Lead), American Association of State Highway and Transportation Officials (AASHTO), National Electrical Manufacturer's Association (NEMA), and SAE International. The project purpose is to develop and publish a CI implementation guide that standardizes the key capabilities and interfaces

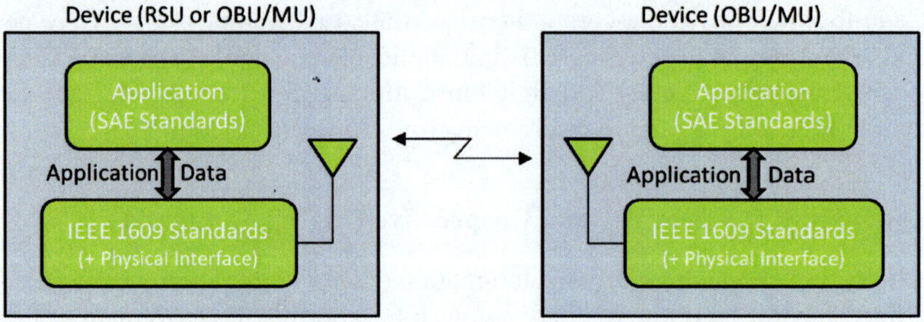

Fig. 5. Connectivity and CDA standards areas (Source: Justin McNew).

for a connected intersection. The guide should address the ambiguities and gaps identified by early deployers and provide enough guidance to generate messages and develop applications for signalized intersections that are truly interoperable across the United States, especially for automated transportation systems.

2.3.3 5GAA Roadmap - Jim Misener, Senior Director, Product Management and Global V2X Ecosystem Lead, Qualcomm Technologies, Inc.

Jim Misener described the high service-level requirements and expected timelines for C-V2X use cases, summarized in Fig. 6. The timeline can be divided into four phases, with increasing level of complexity and technical requirements. From 5GAA, "Cooperative Manoeuvres (via direct communication) and Sensor Sharing to support cooperative perception – both basic functionalities for automated driving, e.g. Highway Pilot – are supported by 5G-V2X. We predict that all new AD vehicles will be equipped with 5G-V2X from 2026, in line with their mass production and entry to the market. Complex interactions between vehicles and VRUs via mobile phones – through both direct (PC5) and network-based (Uu) C-V2X communications – are foreseen to start by 2027."

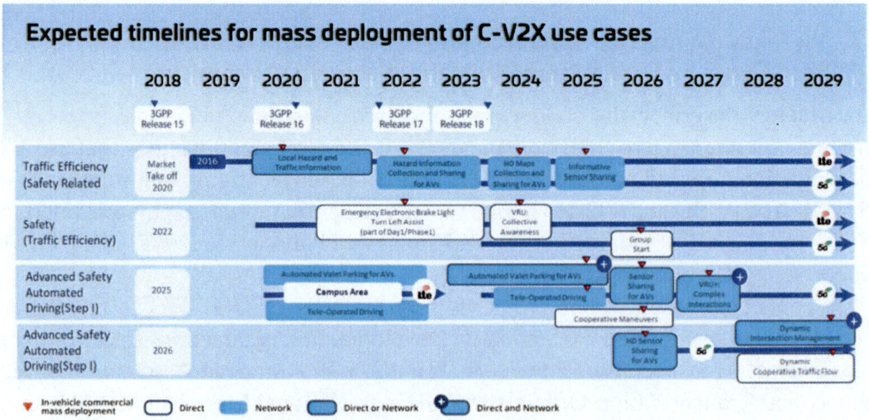

Fig. 6. 5GAA standards roadmap (5GAA 2020).

2.4 Hot Topic 4: Testing and Simulation, Validation and Representativeness

This session explored approaches to simulation verification, and use of simulation for validation and accreditation. Speakers discussed how scenario libraries can be representative, and how they support performance benchmarking. The session was moderated by Chris Schwarz, Director of Engineering and Modeling Research, National Advanced Driving Simulator, University of Iowa.

2.4.1 The National Highway Traffic Safety Administration's (NHTSA) Vehicle Research and Test Center (VRTC) Simulation Research by Scott Schnelle, NHTSA, VRTC

Dr. Scott Schnelle presented the simulation research at the National Highway Traffic Safety Administration's (NHTSA) Vehicle Research and Test Center (VRTC). He reviewed current state of vehicle simulation software, presented methods in generating research scenarios, showed the test results to validate the simulations, and discussed the lessons learned from implementing test procedures in simulation packages. The work started with an initial literature review of simulation frameworks and standards using a 2019 report titled "A Review of Simulation Frameworks and Standards Related to Driving Scenarios". The team selected five example pre-crash scenarios which are described in a 2021 report titled "An Approach for the Selection and Description of Elements Used to Define Driving Scenarios". They also plan to publish their work soon on validating vehicle dynamics model fidelity using subjective analysis with simulation and test track data.

VRTC simulation research team implemented three scenarios in five different simulation software packages. The team performed three kinds of tests:

- Lead Vehicle Interactions - Represented by Automatic Emergency Braking (AEB) test procedures. The team performed 66 tests with varying speed and deceleration rate for the lead vehicle stopped, moving, decelerating scenarios. The validity requirements include principal other vehicle (POV) lane position, velocity, heading, deceleration and timing of relative distance and relative speeds.
- Lane Change Interactions - Represented by Traffic Jam Assist draft test procedures. The team performed 49 cut-in tests with varying speed and cut in range. The team also performed 30 cut-out tests with varying speed and reveal range.
- Vulnerable Road User (VRU) Interactions - Represented by Pedestrian Automatic Emergency Braking (PAEB) draft test procedures. The team performed 96 tests varying prescribed overlap, initial subject vehicle (SV) speeds, VRU constant speed, and environment. The 96 tests include 32 adult, 32 child, 32 bicyclist tests. The validity requirements include VRU speed, heading, position and timing overlap.

The results of the tests show that all five of the simulation packages were eventually able to meet the validity requirements for scenario choreography. While both test modes were able to meet the validity requirements set forth in the test procedures, the tests and the simulations reveal different trends. Most difference are attributed to simulation fidelity. The lessons learned from implementing test procedures in simulation packages is that with only control over inputs to the scenario, it is hard to specify tolerances for subject vehicles with higher-levels of automation.

2.4.2 Euro NCAP and AV Virtual Testing and Assessment by Matthew Avery, Director of Insurance Research, Thatcham Research

Matthew Avery presented how to assess the safety of automated and autonomous vehicles in European New Car Assessment Programme (Euro NCAP) through virtual testing and assessment. To promote safe and balanced automation, the Euro NCAP 2025 Roadmap requires safety grading of L2 and L3 assisted vehicles be provided to consumer. Matthew Avery listed important trends shaping the future of safety testing which include ADAS in complex multi-object scenarios; Holistic (scenario based) assessment of integrated safety functions; Real-world robustness; Population diversity; Flexible, moveable and safe seating; Verify software versions & OTA updates; and Connectivity & cybersecurity.

Euro NCAP has initiatives on virtual testing on crashworthiness & integral safety and ADAS & automated driving. The virtual testing methodology includes industry using ADAS model in simulation and analysis. The results will be used in weighted validation by the test institutes. Matthew Avery presented the idea of combined validation and/or Vehicle in the Loop (VIL) to replace the weighted validation. In the VIL testing, vehicle is driven on test track environment and the "Virtual" platform injects obstacles (cars, pedestrians, etc.) in the ECU algorithm to trigger a vehicle response to the critical event (braking, steering, etc.). The VIL could be a solution to evaluate ADAS & assisted driving functions for Euro NCAP.

2.4.3 Validation Lessons Learned from Aerospace – Edward Chow, Manager, Civil Program Office, NASA Jet Propulsion Laboratory

Edward Chow discussed validation lessons learned from aerospace industry. He pointed out that the roundtrip latency to communicate with a spacecraft on Mars is about 20 min. Because of the long latency to communicate with deep space spacecrafts, with our spacecraft, NASA JPL has been building spacecrafts with autonomous features for a long time. Aerospace industry has been using model-based system engineering for spacecraft design where the design team uses simulation model to design, validate, and test spacecrafts. Spacecrafts have multiple subsystems. The spacecraft bus and instruments could be built by different organizations. So, distributed model-based design and testing between different partners is an important part of the design practice.

Aerospace industry uses the system-level test principle of "test as you fly and fly as you test". The "Test-as-you-fly" requires that ground tests and simulations accurately reflect the planned mission profile, plus margin. Because physics and environment are unpredictable so "Test as you fly" often needs to ensure functional system under infinite conditions. Also, because flight trajectory is often unique, so it may not be possible to "Fly as you test". Validation of autonomous software is an extremely difficult problem. We need to ensure validation won't push software and system engineers to burn-out and won't push the limit of mission budget.

2.4.4 CARMA Evaluation by Philip Azeredo, US DOT VOLPE

Philip Azeredo from the US DOT Volpe Center presented the CARMA Program. CARMA is a Federal Highway Administration (FHWA) program that stimulates collaborative research on the technology, open-source tools, and frameworks designed to improve transportation system mobility, safety, and efficiency. CARMA leads research on Cooperative Driving Automation (CDA), leveraging growth in automation and cooperation to advance transportation systems management and operations strategies. CARMA has been developed as a tool to be able to do research in CDA and facilitate growth in the CDA field. CARMA uses Vehicle to Everything (V2X) as the tool necessary to implement CDA.

The CARMA Evaluation process consists of: CARMA Simulation, CARMA 1tenth, CARMA Testing, and CARMA Analytics. Simulation, 1tenth, and Testing are all ways of testing the CARMA system and CDA applications. CARMA Analytics is a cloud-based data management system used to host data from testing and manipulate and work with it to evaluate the CARMA systems performance. The CARMA Simulation is an everything in the loop (XiL) simulator that uses co-simulation for developing CARMA applications such as CARLA for simulating vehicle dynamics, SUMO for simulating traffic patterns and infrastructure, NS3 for simulating V2X communications. The CARMA 1tenth's goal is to advance the understanding of CDA by using scaled-down C-ADS's. CARMA 1tenth (C1T) can be used to test CARMA features and CDA applications while keeping development and testing costs down. The CARMA Testing includes a fleet of five light vehicle C-ADS's and four heavy truck C-ADS's to enable the testing of CARMA capabilities and CDA applications at full-scale. CARMA Testing occurs at multiple facilities located across the United States.

2.4.5 Validating Safety Critical ADAS/AV Systems by Jace Allen, Director of ADAS/AD Engineering, dSPACE Inc.

Jace Allen from dSPACE Inc. presented end-to-end simulation and validation solutions for vehicle testing, Hardware-in-the-Loop (HIL) testing, data reply, and Software-in-the-Loop (SIL) testing/Cloud. dSPACE provides an open and integrated simulation ecosystem that can support closed loop testing with reusable simulation assets between SIL and HIL. It can enable high realism in simulation and open models via FMI and API, physics-based radar, camera and LiDAR sensor simulation, virtualization of ECUs and bus communication, flexible SuT connection and support of frameworks (ROS, Autoware, etc.), open test tools and interfaces via ASAM xIL-API and Scenarios, Support of standards, e.g., OSI, AUTOSAR, FMI, Open-X. dSPACE can validate functions for autonomous driving and ADAS with realistic vehicle behavior, sensor virtualization, and real-world environment. The sensor model validation can include correct physics behavior based on analytics and high-level simulation results, compare measurement data for simulation system verification, and model and validate sensor specific effects in cooperation with sensor suppliers. dSPACE can generate simulation scenarios from measurement data. This is a highly automated process that can generate scenarios for closed-loop and open-loop SIL/HIL simulation from raw and object lists. It also supports ASAM OpenDRIVE® and OpenSCENARIO® standards.

2.5 Herding Cats

The session concluded with several standards development organizations (SDOs), including from SAE, ISO, IEEE, and ASAM going head-to-head in a "Herding the Cats" panel session to answer the big questions – "Is there any hope of harmonizing AV standards? Can we justify the chaos?" The session explored the roles of key regulatory players, including Country level authorities (Ministries of transport), State or province/local government authorities (DMVs, DOTs), Federal level (e.g. NHTSA), UNECE WP.29 (World Forum for harmonization of Vehicle Regulations), and World Economic Forum (Influencer).

Topics where there is seemingly overlap or conflict were discussed. For example, the term "scenarios" is used differently across standards:

Scenario - description of the temporal relationship between several scenes in a sequence of scenes, with goals and values within a specified situation, influenced by actions and events.

Scenario - Sequential description of the scenes integrated with the ADS(s)/subject vehicle(s), and its/their interactions in the process of performing a certain dynamic driving task(s).

Scenario - A description of the temporal development through several consecutive scenes in a sequence of scenes.

Scenario - A scenario describes the traffic, infrastructure and environment (including e.g. weather and lighting conditions) for the simulation and consists of a sequence of scenes. It is limited in terms of time and space.

Another controversial topic is the lexicon and ontology for ODD. For example, AVSC is missing V2V (no 802.11p) and Interference, PAS1883 has more detailed list on the communications types, AVSC focusses on end-to-end connectivity (fleet management), and different terms are used for signal strength (PAS1883) versus Obstructions (AVSC).

Panelists discussed ways that coordination and deconfliction are being handled, including formal agreements between SAE and IEEE and UL, as well as common participants across SAE, ISO and UNECE efforts. All participants agreed on the need for a shared taxonomy to clearly identify core goals of various activities, and to more clearly position the outputs.

Aurora presented some work they had conducted to better understand and organize the "chaos" in the standards space (see Fig. 7).

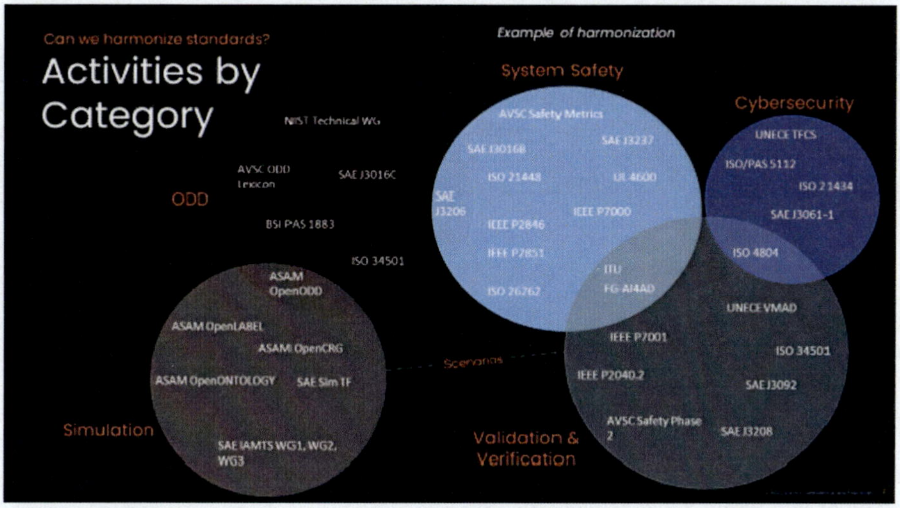

Fig. 7. Mapping of published or current standards activities by topic area (Source: Aurora).

3 Conclusions

There is a growing list of AV standards publications and activities at SAE, ISO, ASAM, JAMA, UL, MUTCD, UNECE, among others. SAE CAVSource is currently tracking 129 international AV-related standards activities and needs in a roadmap. So what's real and what's next?

We focused on 4 hot topics: "Safety Assurance Demonstration", "Physical Infrastructure", "Connectivity and Cooperative Driving Automation", and "Testing and Simulation, Validation and Representativeness"; which are the subject of many current standards and regulatory activity.

We attempted to herd the cats and answer the question "what's next?". There were some areas we achieved clarity and clear recommendations, and other areas we managed to justify the chaos.

For Safety Assurance Demonstration standards, there is co-existing and overlapping, but increasingly coordinated, efforts at SAE, ISO, ASAM, UL, UNECE, especially on topics related to metrics, principles, usage of on-road, track, and simulation testing, and scenario-based testing.

For Physical Infrastructure standards, we are anticipating a series of SAE Edge Reports, CSA's CAV Framework publication, and FHWA research. A common theme was that Infrastructure Owner Operators (IOOs) need more information regarding what is needed to help with localization, lane management and dynamic traffic control elements (signals, work zones, rail crossings).

For Connectivity and Cooperative Driving Automation standards, cooperation between vehicle and infrastructure can share the load of executing ADS use cases, e.g., communications and computational load. We covered recent advances in cooperative intersections, traffic management, and perception.

For Testing and Simulation, Validation and Representativeness, it is arguably impossible for a single database to contain all ADS scenarios. The fast pace at which new ADS technology comes to market makes it difficult to maintain state-of-the-art insights from ADS simulation. Non-collaborative approaches to ADS simulation are perhaps more expensive than collaborative approaches. Simulation standards are difficult to define because stakeholders have different needs and priorities, making collaborating even more challenging.

4 Next Steps

Session organizers and participants, many of whom are leaders in the standards space, offered to take these discussions forward in standards development efforts. Presentations have since been given at standards committee meetings summarizing the outcomes of these sessions. Public and private sector stakeholders stand to benefit from understanding the latest in standards, and how they enable the deployment of AVs. The following key outcomes and research needs were identified:

For Safety Assurance Demonstration standards, common definitions, principles, and precise language are key enablers to defining safety goals and methods. We need continued coordination among many activities. To enable the use of simulation, we need best practices for usage of simulation and validation / benchmarking.

For Physical Infrastructure standards, we need a common language for discussing infrastructure elements and needs, including isolating physical and digital elements. There are opportunities for AVs to share dynamic data, including weather and degradation / maintenance related issues to IOOs.

For Connectivity and Cooperative Driving Automation standards, we need standards that define CDA use cases to be fed into interoperable testing platforms that enable cooperative testing; and related, we need coordination needed between automotive and infrastructure communities.

For Testing and Simulation, Validation and Representativeness, ADS co-simulations and collaborations will help the industry set a baseline threshold for safety. (Baselines set by experts is important to consumers.). The ADS simulation and validation discipline could greatly benefit in pursuing more efforts for database harmonization and compatibility. Stakeholders could communicate the minimum expectations of each other's automated systems for ADS simulation collaboration to be realized. More could be learned about vehicle dynamics and ADS model fidelity. The aerospace industry may be able to provide suggestions on overcoming barriers to ADS simulation collaboration. The establishment of ADS simulation standards could improve ADS simulation collaboration. Different sectors could consider using ADS co-simulation to develop a baseline for ADS safety.

References

5GAA: A visionary roadmap for advanced driving use cases, connectivity technologies, and radio spectrum needs, 9 September 2020. https://5gaa.org/wp-content/uploads/2020/09/A-Visionary-Roadmap-for-Advanced-Driving-Use-Cases-Connectivity-Technologies-and-Radio-Spectrum-Needs.pdf

ASAM: ASAM OpenX Standardization for validation and verification of autonomous driving, October 2021. https://www.asam.net/index.php?eID=dumpFile&t=f&f=4489&token=0bef3540f5d247e6352acdf2615c361dc1cee7e6

CSA: Connected and automated vehicle technologies – insights for codes and standards in Canada, June 2020. https://www.csagroup.org/wp-content/uploads/CSA-Group-Research-Connected-Automated-Vehicle-Technology.pdf

CSA: Physical and digital infrastructure for connected and automated vehicles (CAV) – code framework, October 2021, https://www.csagroup.org/wp-content/uploads/CSA-Group-Research-Physical-and-Digital-Infrastructure-for-Connected-and-Automated-Vehicles-_CAV_.pdf

ISO: ISO TC22-SC32-WG13, ISO TS 5083, Safetronic 2021, November 2021. https://safetronic.fraunhofer.de/wp-content/uploads/2021/11/Nov17-UpdatesStandarization-Fuerst.pdf

SAE CAV Source: August (2021a). https://www.sae.org/highlights/cav-source

SAE Edge Report: Accessed August (2021b). https://www.sae.org/publications/edge-research-reports

Six Years of Reading the Road Ahead: Supporting Roadway Automation with Traffic Control Devices

Scott O. Kuznicki[1,2](✉)

[1] Modern Traffic Consultants, Aberdeen, USA
sk@moderntrafficconsultants.com
[2] Discovery Institute, 208 Columbia Street,
Seattle, Washington 98104, USA

Abstract. Improvements in the safety performance and efficiency of the roadway transportation system are sought and realized through the deployment of Automated Driver Assistance Systems and higher-level vehicle automation systems. The consistent function of these systems depends largely on the ability of the vehicle sensors to accurately detect the roadway environment. This environment includes pavement markings and roadside delineation, the primary local offline source of information on roadway alignment. The Reading the Road Ahead workshops at the Transportation Research Board's Automated Vehicles Symposium (AVS, 2016 through 2020) and Automated Road Transportation Symposium (ARTS, 2021) were convened to provide a platform for understanding the interactions of machine vision systems with traffic control devices, featuring expert presentations in the fields of machine vision, human factors, traffic engineering, and transportation safety performance.

Keywords: Vehicle automation · Traffic control devices · Machine vision · Human factors · Traffic engineering · Transportation safety performance

1 Introduction

The development of automated driving systems (ADS) has progressed concomitant with advancements in sensor technologies. Sensor systems on vehicles perform a function similar to the sensory organs and systems of the human body, including perception, detection, and information transmission. Like human sensory organs, even the most advanced microelectronics can fail to adequately detect and interpret information. This chapter addresses the general approach and reported findings of the Reading the Road Ahead workshops over a six-year period, addressing machine vision system interactions with traffic control devices in the context of the ongoing need to address the performance of traffic control devices relative to the needs of the human driver.

G. Meyer and S. Beiker (Eds.): ARTSymposium 2021, LNMOB, pp. 140–148, 2023.
https://doi.org/10.1007/978-3-031-11112-9_11

1.1 Reading the Road Ahead

These workshops build on extensive research work related to visibility of traffic control devices, which is the most basic performance measurement of any sign or marking. Human drivers and machine drivers are expected to share roads for the next two decades; thus, the workshops frame the conversation in the context of relating performance to the capabilities and limitations of human drivers. Correlating these aspects of human performance with design and operational characteristics of the roadway environment provides a basis for understanding how machine vision systems will be integrated into the existing roadway system as a user while improving and augmenting the performance of human vehicle operators. Understanding machine vision system performance through a framework of capabilities and limitations (the basis of human factors science) allows researchers to identify gaps in performance for both human drivers and machine vision systems.

1.2 The Driving Task

Human factors science is applied to transportation engineering primarily through the study of human behavior and road user interactions. These interactions include those occurring between the road user and the vehicle (trucks, buses, cars, motorcycles, bicycles) either as an operator or as a pedestrian or passenger, and those occurring between the road users and the roadway environment. The roadway environment includes the entire roadway within the sensory perception range and is typically centered around the primary sensory input for driving, within the cone of vision. Through this primary field of view, road users find various roadside appurtenances such as illumination and utility poles, other vehicles and people, traffic signals, signing, and pavement markings and delineation.

The human driver must detect, identify, and process information from the roadway environment. This process incurs workload on the human sensory and cognitive systems across three general task areas, classified as navigation, guidance, and operation. This process is illustrated as the human factors triangle in Fig. 1, which also illustrates corresponding task areas related to machine vision systems. Navigation involves visualizing

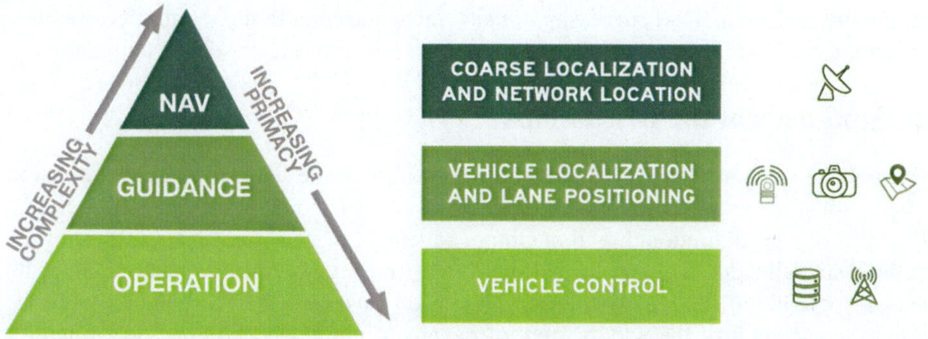

Fig. 1. Relating the Task Workload/Primacy Triangle and Machine Vision, from slides presented by Scott O. Kuznicki, P.E. to the ITS World Congress 2019, Singapore

the vehicle's location along a desired route or within a network, e.g., the process of finding the way from one point to the next. The process of navigation influences lane selection and turning movement activities. Guidance involves the process of keeping the vehicle within a lane and positioning it along the roadway, a process that largely depends on the presence of pavement markings or roadway edges that are visible. Vehicle operation, which must be performed consistently for the duration of the trip, is the least complex task and involves physical operation of the controls of the vehicle.

For the human driver, too much attention paid to one component of the driving task can lead to performance degradation. Concentrating on a map (navigation) can lead to a diversion of attention from activities related to the guidance task (looking down the road, maintaining following distance), such that inattention could contribute to a crash. Correspondingly, the demand for attention from a single task such as guidance is often related to the lack of adequate information.

The most prescient example of task saturation due to a lack of information is the workload incurred by a driver during precipitation events when the road is wet and pavement marking visibility is subsequently reduced. Additional attention must be paid to detecting the pavement markings and most drivers reduce speed to compensate for the additional workload. The presence of retroreflective raised pavement markers (RRPMs) provides additional information that is readily detectable, reducing the workload associated with guidance. (RRPM's are typically spaced at intervals between 40 and 100 feet (approximately 12 m to 30 m) in the United States, but used in limited circumstances in states where snow is common.) While an increase in speed is expected, the overall reduction of adverse crash outcomes and increase in driver confidence and lane-keeping precision indicates the positive effect of reducing workload and limiting task saturation, which can also provide for more effective operation of machine vision systems, particularly where contrast and reflection issues occur.

This workload reduction is affected largely by the visibility and differentiability of the markings (color, pattern, width, application) and aided for both humans and machines when consistency is provided. Consistency in the application of marking patterns and colors to specific scenarios is critical for creating expectations based on logic models, the foundation of computational science.

The key finding of the Reading the Road Ahead workshops is that consistency in application, differentiability between applications, and visibility under a wide variety of conditions is fundamentally necessary for pavement markings to support the performance of human drivers and machine vision systems across all levels of vehicle automation.

2 Summary of the Discussions

Focused on the performance of pavement markings in the context of machine vision interactions, the workshops have addressed issues related to sensor capabilities and limitations, high-definition digital mapping, pavement marking materials and visibility, vehicle spatialization using pavement markings and high-definition mapping, and the use of logic-based models to address the use of pavement marking colors, patterns, width. On account of the wide variety of pavement marking condition and variability in implementation, the workshop has also addressed international research related to Operational Design Domain selection and road rating systems for self-driving transport.

During the course of the workshops, a variety of industry and academic research perspectives were presented by contributors. These contributions addressed the following general topics, with selected speakers highlighted.

- State of the practice; sensor systems architecture and capabilities
  Robert Seidl, Leo McCloskey, Richard Worl, Scott Kuznicki, Ethan Sorrelgreen
- Digital mapping, data collection, and classification
  Monali Shah, Jennifer Carter, Ro Gupta, Siddartha Khastgir, John Corbin
- Pavement marking and sensor performance
  Phil Magney, Adam Pike, Ethan Sorrelgreen, Ken Smith, Tammy Meehan Russell, Doug Campbell, Kevin Sylvester, Paul Carlson
- Field applications of machine vision integrations
  Ross Sheckler, Peter Kozinski, Angelos Amditis, Doug McClanahan, Brian Simi
- Process and oversight architecture and organization
  Tom Alkim, Sivakumar Rathinam, John Corbin, Jaap Vreeswijk

The author, along with TRB and AUVSI, gratefully acknowledge the contributions of Ken Smith, Paul Carlson, John Corbin, Robert Dingess, and John Obenberger, along with all of the organizations' key partners at the U.S. Department of Transportation and transportation infrastructure entities throughout the globe.

Workshop outcomes over the six-year period have included recommendations related to technical and policy issues. One conclusion of the workshop's panelists and participants is that increased resources must be devoted to improving the consistency of pavement marking visibility and application in a systematic framework, particularly for government-operated roadways. Each year, this author presented photographic summary of traffic signing, pavement markings, delineation, and work zone traffic control devices. These surveys indicated that significant progress remains in ensuring consistency of application of marking patterns, colors, and widths. Additionally, all presenters addressed monitoring of TCD conditions as a means of managing maintenance to achieve a state of good repair. Resources applied to roadway infrastructure maintenance have been demonstrated to improve the uniform application in pavement markings, both in logical consistency and in physical condition. Roadway systems operated under concession or by non-government entities generally achieve superior results due to a focus on contract and customer-centric outcomes, wherein quality management for maintenance and operations is emphasized.

2.1 Context and Scope

Research related to pavement marking visibility and performance is maturing and the factors contributing to performance degradation of machine vision systems are understood to be related to pavement marking pattern, materials, contrast, ambient conditions, and the capabilities and limitations of the machine vision systems. Conventional means of measuring visibility (pavement marking retroreflectivity) are undertaken under controlled conditions and the wide variety of field conditions necessitates in-field measurement of performance, an issue addressed by researchers at the Texas Transportation Institute and VSI Labs. This research has indicated a need to better understand how to measure, report, and prepare responses to changes in pavement marking visibility due to ambient conditions. A straightforward approach of assessing the distance over which markings can be seen is readily supplemented by alternative *validation* of spatialization and the use of near-live mapping to facilitate control transitions to the human driver from ADAS or HAV operations.

Road Assessment Systems for ADAS and HAV Operations

Classification of physical infrastructure elements is a challenge that is exacerbated by inconsistent applications of traffic control devices, variations in standards between agencies, even within countries, and shrinking funding for public agencies on account of competing non-highway purposes. Development of a classification system that can be applied in any context is difficult. Across multiple agencies, not all harmonization for applications are followed (standards, guidance, options, and regional practices), such that the meanings of markings in one location may differ from the meaning of markings in another location.

All of these roadway elements serve a purpose even as performance limitations are evident due to variations in application, maintenance, and individual conditions. Thus, the classification of the various modes in which road users and machines interact is variable even when ambient conditions and geometric design are similar enough that no performance difference is expected.

Differentiation between different styles of pavement markings (broken lines, dotted lines, and dotted extensions, for example) can be quantified through an analysis of the output of machine vision system perception and identification. Simulation can then be used to validate and support pavement marking patterns being recognized and categorized by machine vision systems receiving inputs from visible light and LIDAR sensors and interacting with HD mapping systems. Improved pavement marking technologies are addressing the needs of machine vision systems, particularly in the area of contrast in low-light and obscured-atmosphere conditions (1).

This output data from machine vision sensor reading of the roadway can also be supplemented by huge datasets from naturalistic studies, which continue to be a major asset in understanding vehicle-road-user interactions and classifying a wide variety of incidents, including non-trivial lane departures and lane-keeping failures. Collection of this data in ADAS-equipped and HAV-compliant vehicles will further provide researchers and technologists with the ability to effectively crowdsource large amounts of data related to how well machines read the road ahead and the interactions of pavement markings with machine-based driver assistance and self-driving systems.

The development of a road assessment system has been addressed in research conducted at Warwick University and through the TRB activities associated with the International Symposium on Traffic Signs and Pavement Markings, held in Zagreb, Croatia, October 2019. Such road assessment systems have the potential to assign ratings to various aspects of roadway environment for fixed, transient, recurrent, and degrading qualities, including those related to pavement markings and delineation and associated with reliability grades incurred as a result of asset and weather-related maintenance activities (Fig. 2).

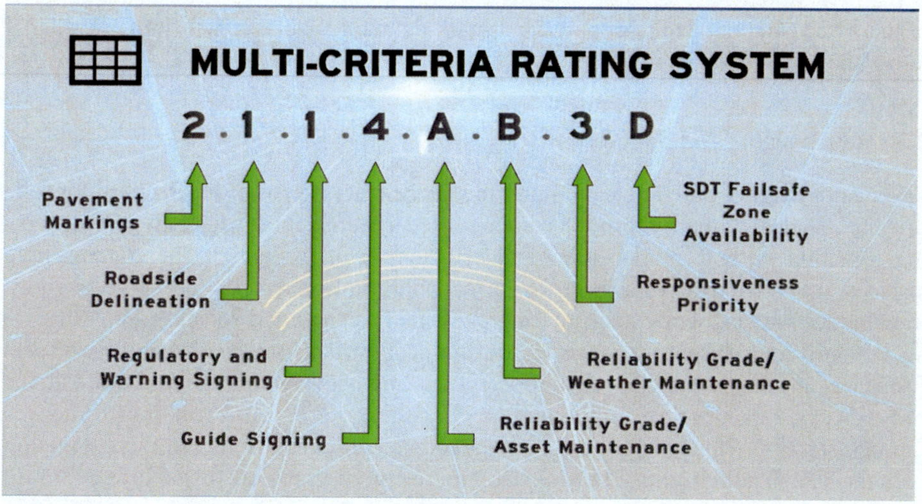

Fig. 2. Potential multi-criteria road assessment output, from slides presented by Scott O. Kuznicki, P.E. to the ITS World Congress 2019, Singapore

2.2 Recent Research and Trends

The rollout of machine driver integration from companies such as Aurora and Waymo stands to bring Level 4 and Level 5 automation to a wide variety of vehicles, including Class 8 trucks. Meanwhile, continued evolution of ADAS capabilities within the Level 1 and 2 framework will increase the risk of task deprecation due to improved performance. The likelihood of failed control transfers will also possibly increase, despite the requirement for constant human driver supervision of Level 1 and Level 2 operations. Improving the primary means of lateral spatialization will reduce the risk of occurrence for these control transfers, particularly in marginal conditions.

Ongoing research will continue to assess the visibility performance of pavement markings under conditions where markings are obscured by water and contrast performance is inhibited by ambient lighting and ambient conditions affecting machine vision systems. This research stands to improve the correlation between pavement marking retroreflectivity measurements being undertaken by road agencies and the expected performance of various machine vision systems. These correlations will also be applicable to ODD architecture and selection.

The development of federal, state, and road agency policies does not necessarily mean that those policies will be implemented in a timely manner or that they will be implemented with a degree of consistency that can provide confidence for AV machine vision systems.

Defining ODD for motorways in particular will require multi-faceted definitions of ODD for various conditions and machine vision capabilities, with extensive work required related to taxonomy and identifying boundary conditions for users and environment.

FHWA's updates to the Manual on Uniform Traffic Control Devices are anticipated to play an essential role in providing a framework for continued progress and justification of increased funding levels for pavement marking maintenance, which supports ADAS today and HAV in the near future.

2.3 New Insights and Suggestions

Pavement markings and delineation provide the primary means of localization for self-driving vehicles, primarily through camera-based systems; it is critical that these markings function for both human users and machine vision systems in the coming three decades, supported by pavement surfacing technology advancements that serve to reduce interference with camera systems during inclement weather and enhance durability.

Attention must be paid to consistency of application, consistency of visibility and performance, and consistency between jurisdictions and between functional classifications of roadways. In support of improving the consistency of pavement marking practices through safety-oriented programs, the workshop's participants suggested coordination with the TRB Traffic Control Devices Committee on two projects, listed here.

- Survey of Practice: State DOT Pavement Marking Maintenance Practices and Funding Mechanisms
- Road Safety Research: Quantifying ADAS-equipped Vehicles' Contribution to Road Safety Across Multiple Operating Design Domains

3 Conclusions

Over the course of six years, workshop panelists and participants reported conclusions related to three areas of technological development addressed by the Reading the Road Ahead workshops. These three areas include pavement marking performance, machine vision capabilities, and supporting technologies. The use of road assessment systems and ODD correlation can assist researchers in identifying the relationships between pavement marking performance (driven by materials and implementation and variable according to ambient conditions) and machine vision system capabilities and limitations. These relationships are inherently variable (due to the changing angle of the sun or even position within a lane) and cannot be classified by ODD alone.

In 2020 and 2021, researchers suggested that pavement markings will remain essential for both human drivers and machine drivers and that the maintenance of pavement markings and roadway surfaces will remain critical to the successful operation of ADAS and HAV systems. Allocation of resources to road pavement maintenance and pavement marking activities should therefore become a priority for road infrastructure owners and operators. Managing the surface of the pavement as a multi-faceted system and the primary means of providing adequate information to ADAS and HAV systems, facilitating high reliability across variety of conditions.

The preliminary findings of the 2017 to 2019 workshops were presented at the first International Symposium on Traffic Signs & Pavement Markings in Zagreb, Croatia. Four characteristics of pavement marking systems were identified as critical aspects of performance related to the interaction between machine vision systems and pavement marking and delineation systems, as displayed in Table 1.

Table 1. Critical criteria for assessing pavement marking system suitability for a multi-user environment, organized by logical modes and functional modes

CONSISTENCY	Use of marking and delineation patterns across various applications is subjected to a logical similarity test such that there is no mis-match of marking types and geometric scenarios so that discrete scenarios can be related to discrete patterns of TCDs
DIFFERENTIABILITY	Markings of various types are used in such a way that each marking type is distinguishable from others and that individual geometric scenarios can be distinguished based on markings alone
PERFORMANCE	Visibility, discernability, and durability of pavement markings is measured and managed to meet minimum performance criteria applicable to both human and machine road users
MAINTENANCE	Maintenance of markings, the roadway surface, and the roadway environment ensures visibility of traffic control devices and preserves contrast across a variety of ambient conditions and throughout the duration of the facility's operational life

These findings are organized according to the architecture proposed in previous work, which addresses the logical consistency requirements of effective traffic control devices (2) with a particular focus on pavement marking patterns.

ADAS implementations are saving lives and the consistent delivery of ADAS operations requires that ongoing surface maintenance must be systematically funded to reflect the road safety contribution made by pavement markings and well-maintained pavement surfaces.

4 Next Steps

Workshop participants generally agreed in 2021 that international research efforts seeking to understand the infrastructure needs of automated driving systems must continue in order to promote safety for all road users, there exists a continuing shrinking assurance of sustained funding from governments. In some jurisdictions, government funding for the maintenance of preservation of pavement markings is insufficient to assure the adequate performance of markings for both human drivers and machine vision systems. Advocacy for increased funding related to traffic operations and pavement maintenance may prove to be a stopgap measure and the research highlighted by the workshop will demonstrate that these future expenditures are essential if transportation safety is to be maintained.

Looking beyond governments to industry associations and infrastructure investors holds a great deal of promise, as these technologies continue to create economic opportunities across multiple sectors and improvements can be easily monetized. Industry associations play a crucial role in harmonizing policy approaches and create investor confidence as the associations will seek to prioritize safety and convenience for all travelers, driving continued investment in motorways, the most robust components of the transportation system.

Private equity firms that invest in the development of self-driving transportation systems (from machine vision equipment to full-stack software and hardware integration) are incentivized to better understand infrastructure and the needs related to automated driving. The potential for private equity investment supporting infrastructure development has been demonstrated in Europe, the Middle East, and Southeast Asia, and in some parts of North America. These facilities, which are managed for the benefit of the customers (travelers) and investors, yet open to public travel and often held in the public trust outside of the auspices of elected offices and administrative agencies, generally demonstrate more consistent attention to the quality and performance of pavement markings and delineation. Ease of driving is considered key to the customer experience and safety performance by many road operators. As machine vision systems become the dominant user of these facilities, motorways in particular, adaptation to these new drivers and the experience of the customers in the vehicle will continue to drive prioritization of maintenance and preservation activities. The quality of these facilities will attract additional investment in expansion supporting an improved travel experience and price stability.

The inconsistency of pavement markings across a wide variety of roadway functional classifications and jurisdictions necessitates the preparation and dissemination anticipatory information for ADS, particularly for SAE Level 3 operations, where a pre-emptive machine driver handoff to the human occupant is essential. Such information could be readily handled by an industry-wide clearinghouse for infrastructure data collection and classification topologies may best be managed by means of a new industry association focused on road assessment systems and practice assessments that target improved consistency in the application and maintenance of pavement markings.

References

1. Pike, A.D.: LiDAR detection and sensor fusion opportunities for pavement markings. In: Proceedings of the Automated Vehicles Symposium, Orlando, Florida (2019)
2. Kuznicki, S., Katz, B.: Designing for consistency: matching applications to scenarios in the use of traffic control devices/pavement markings. In: Proceedings of the 5th International Symposium on Highway Geometric Design, Vancouver, BC, Canada (2015)

Remote Support for Automated Vehicle Operations

Jaap Vreeswijk[1(✉)], Azra Habibovic[2], Olav Madland[3], and Floris Hooft[1]

[1] MAP Traffic Management, Van Deventerlaan 20, 3528 AE Utrecht, The Netherlands
{jaap.vreeswijk,floris.hooft}@maptm.nl
[2] Scania, 417 56, Göteborg, Sweden
azra.habibovic@scania.com
[3] Applied Autonomy, Kirkegårdsveien 45, 3616 Kongsberg, Norway
olav.madland@appliedautonomy.no

Abstract. What is remote support, why is needed and who is the service owner? Remote support comes in different shapes and forms. Typically, a remote human operator provides instructions, permission or waypoints to the vehicle, or remotely drive it. However, the purpose and tasks of the operator are very diverse for different modes and environments: ports and yards, (on-demand) taxi services and low-speed passenger shuttles, long-haul commercial trucks, road-based last-mile/middle-mile delivery vehicles, sidewalk delivery robots, privately owned vehicles, autonomous vessels and air mobility, on confined areas, segregated and shared infrastructure. Moreover, there is an intuitive synergy with road traffic management and fleet management services, and eventually multi-domain integrated operations management centers. This chapter aims to increase awareness and understanding of remote support by sharing experiences and achievements alongside discussion of technological requirements, operational aspects and future research needs.

Keywords: Remote support · Control room · Tele-operation · Automated vehicle · Operator

1 Introduction

There are many situations which automated driving systems (ADS) cannot handle or require many years of development. Over the past years remote support has received an increasing amount of attention as now and in the foreseeable future, a safe and comfortable autonomous transport service in mixed traffic without a steward on board (SAE-level 4), is expected to rely on some form of remote control. Typically, remote support involves a remote (human) operator providing instructions, permissions or waypoints to the vehicle, or remotely driving it. It is sometimes referred to as tele-operation, remote supervision, vehicle operations management, command center, control room or remote monitoring and control.

Remote support is considered most useful when the vehicle encounters unknown situations or when illegal actions are required. Typically it means that the vehicle cannot

G. Meyer and S. Beiker (Eds.): ARTSymposium 2021, LNMOB, pp. 149–164, 2023.
https://doi.org/10.1007/978-3-031-11112-9_12

execute one or more of its driving tasks and needs support in either environmental awareness, decision-making or actuation in order to proceed. In addition, remote support enables operations, ensure safety and increases public acceptance.

The purpose and tasks of the operator can be very diverse for different modes and environments, ranging from confined areas for cargo movements to passenger vehicles on public roads. The purpose of this breakout session at ARTS2021 was to increase awareness and understanding of remote support: what is remote support, why is needed and who is the service owner? Another objective was to identify remote support functionality and technological requirements based on real-life concepts of operations. Finally, when looking at wider deployment of remote support synergies with road traffic management and fleet management were discussed, ultimately leading to multi-domain integrated operations management centers.

Remote control means that an automated vehicle is controlled by a human operator over mobile radio networks. This is particularly relevant in case of undefined, unexpected or exceptional (traffic) situations, which the vehicle is not capable to handle. A human operator assists to solve the situation with the support of software tools, while looking after the passengers and informing them if needed. Broadly speaking four levels of remote support can be distinguished:

1. **No assist**: perception, decision-making or actuation are fully executed by the automated vehicle which has primary safety and mission responsibility.
2. **Remote assist**: the automated vehicle has primary safety and mission responsibility and a remote human operator provides instructions, waypoints, missions or permissions as needed.
3. **Remote control**: temporary full operational control typically to resolve a situation, also known as remote human driving. All perception and actuation tasks are executed by the human remote operator which has the primary safety responsibility.
4. **Shared control**: remote human driving while the vehicle controls the on-board crash avoidance systems, or remote assessment of a situation and providing concrete operational guidance recommendations which are executed by the vehicle (e.g. environmental awareness or trajectory). In both cases there is a shared safety responsibility.

As an example: an automated vehicle encounters an object on the road on a busy street. The vehicle comes to a stop and request support from the remote support center. A human operator receives in real-time the video and sensor feeds from the automated vehicle to understand the full context of the situation. The operator provides guidance for a clear and safe path around the object. Finally the vehicle determines when the time is safe to follow the recommended path. Figure 1 illustrates this flow of events.

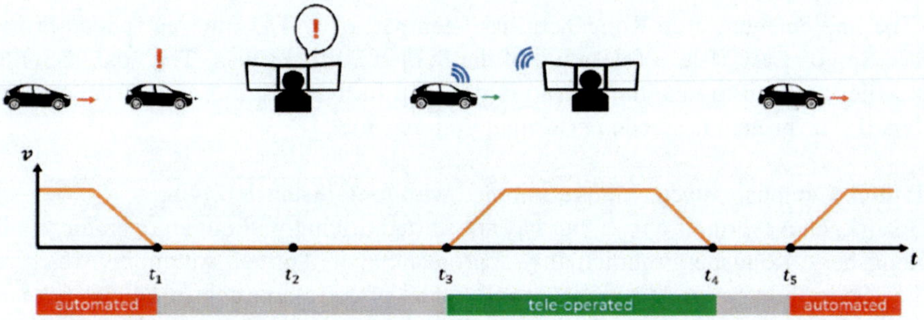

Fig. 1. Timeline hand-over vehicle control and remote control (Source: 5GCroCro).

2 Summary of Presentations

The breakout session consisted of three panel sessions. Panel session A set the scene by providing an overview of remote support for automated vehicle operations with different examples of concepts of operations and the solutions developed. Panel session B addressed features and technological aspects related to remote support, such as: connectivity and communication, vehicle and roadside surveillance equipment, standardization, scalability, cybersecurity, etc. In the third and final session the panelists discussed implementation and research needs for remote support and discussed synergies with other roles and responsibilities in the operational domain. The following sections highlight two speaker presentations and summarize the other presentations as well as main takeaways, open issues and topics for future research collected during the break-out session.

2.1 Experience and End User Feedback from Operation of Public Transport Without Safety Driver 2021

This section presents the experience from Kongsberg where a remote controlled autonomous shuttle in on-demand operations has been implemented in mixed traffic from September 2021.

2.1.1 Motivation and Objectives

The main motivation for implementing remote controlled shuttles without a safety driver on board is sustainability. To reach an attractive solution, the solution must provide a number of important elements. These are:

- Efficiency - higher speed, fewer unplanned stops;
- Attractiveness - drive where people want to travel;
- Economy - drive without an operator on board, cheaper vehicle;
- Trust - documented safety, safe and good customer experience;
- Sustainable - sharing solution for efficient vehicle utilization.

The implementation in Kongsberg has been part of an EU Interreg funded project named Sohjoa Last Mile, a follow-up of the Sohjoa Baltic project. The goal of Sohjoa Last Mile has been to take the operation one step further: driving without an operator on board. The project has been performed in three cities:

- Tallinn: Campus, Auvetech bus combined with long-distance driving
- Gdansk: Graveyard, as above, but has not started driving without an operator yet
- Kongsberg: Kongsberg municipality is project owner, Applied Autonomy was given the responsibility as project manager and vendor of the remote control solution, the on-demand software as well as safety responsibility for the operations. Applied Autonomy decided to use an EasyMile EZ10 bus with remote monitoring and operation. Regular bus drivers have been hired from Vy bus by Applied Autonomy for the remote control center. This means that Applied Autonomy delivered a turnkey solution to Kongsberg municipality and their partners.

2.1.2 Important Details for the Kongsberg Pilot

The mission of the project has been to test self-driving technology with remote-controlled operation of vehicles in the most complex situations possible, testing the capacity limits of the technology without compromising safety and implementing a pilot where the complexity could be increased based on the experiences gained along the way. The mission of the project has also been to test the technology's capacity in relation to public transport's need for flexibility/on-demand transport and identify any barriers in the technology and collaborative systems in relation to the desired flexibility and cost. Sohjoa Last Mile also aimed to gain feedback from users' experiences with regards to usefulness, safety and cost when driving completely without a safety operator on board.

The implementation was performed in accordance with the Norwegian law permitting operation of self-driving vehicles without a driver on board. The Norwegian road authorities regulate both public roads and private roads with one common set of laws and regulations. This means that there are no special exemptions for private roads, even to your own house or farm.

The service is in operation in Kongsberg technology park with 5200 employees inside the park (production, office, goods reception). There are different internal road users (pedestrians, cars, trucks, forklifts…), visitors with large trucks, maintenance workers and guests visiting the park for regular business meetings. The speed limit in the park is 30 km/h. The service has been implemented with 10 bus stops along a 1.5 km long road network (Fig. 2).

All operations in Norway with autonomous vehicles have to be permitted by the National Road Authorities. The applications for permits consist of documentation of the vehicle, the risks assessment for the operations, safety organization, safety procedures and GDPR routines. The approved application and implementation was divided into four phases:

1. Operator on board;
2. Operator on board and operator in a control center;
3. Hidden operator on board and operator in a control center;

Fig. 2. The autonomous shuttle, driving without any adaptation in infrastructure such as road/guiding markings

4. Operator in a control center.

There has been no adaptation in the infrastructure with the exception of information about duty to give way at some intersections. Training of operators has been carried out by Applied Autonomy in collaboration with EasyMile.

Sohjoa Last Mile in Kongsberg has been focused on the trust that has to be built gradually in each phase. The service ran in 100-50-50 h in phases 1–3 and recorded situations relevant to safety and accessibility in a tool which gives full transparency to all involved stakeholders in real time.

The remote control room has continuous situation monitoring using the bus' cameras and telemetry. Audio communication between the bus can be initiated by passengers and the control center. Digital documentation of situations that occur on the route is up to date. There has been a gradual decrease in the number of interventions by the bus operator, and an increase in takeovers by the control center operator.

In phase four, on the random routes ordered by the public via the on-demand solution, the field operator is only in the bus if there has been an emergency stop, or there is an obstacle that must be circumvented manually.

There are to date 99 passengers who have used the shuttle in phase 4. No users are unhappy or even slightly unhappy with the service without a safety driver on board. The operations take place on working days from 07:30 to 14:15.

There have been 27 interventions on the ground needed which represented situations that could not be solved through the control. These were caused by vehicles and goods parked in the areas where the bus operates.

Experiences so Far:

- Fortunately no serious incidents have occurred;
- It turned out to be challenging to have a clear route without e.g. delivery of goods, craftsmen parking, etc.;
- With better camera coverage in blind spots for validation of the situation picture, the overtaking function could have been used in such cases (next step?);
- Marking the line in exposed places could lead to fewer parking obstructions?
- Hidden gift cards in the bus turned out to positively affect people to get to try out the service;
- A survey in the ordering app is available.

Lessons Learnt:

- Efficiency - higher speed, fewer unplanned stops;
- Attractiveness - drive where people want to travel (New booking solution established);
- Economy - drive without an operator on board, cheaper vehicle (No-Op in operation);
- Trust - documented security, safe and good customer experience (Approval process and established control center with customer contact);
- Sustainable - sharing solution for efficient utilization;
- What must the control center support have to take No-Op further.

2.2 Human Factors in Remote Operation of Automated Vehicles

Current trends indicate that remote operation will play an important role for deployment of automated vehicles, and that human remote operators are expected to be responsible for a variety of tasks. For instance, Argo.ai want to use remote guidance, as they refer to remote operation, in "a selected group of particularly challenging conditions, when the self-driving system is unable to make a requisite decision, or requires additional guidance to do so" [1]. However, Argo.ai emphasize that guidance does not include remote driving of the vehicle. This coincides with the difference between the use of "remote operation" vs. "remote support". Similarly, Cruise envision remote operators to provide "bread crumbs for decisions on whether to reroute and how to get around a blocker" without actually driving the vehicle [2]. Waymo points out also that the final decision is always with the automated driving system: "it can call on our Fleet Response specialists to provide advice on what route might be better or more efficient and then take that input, combine it with the information it has from the onboard map and what it's seeing in real time via the sensors, and choose the best way to proceed." [3]. Nuro, on the other hand, state that they will use remote operators "to remotely monitor a vehicle and take over if required…in certain complex situations such as a partial road closure due to construction or an accident [4]. Similarly, Einride envision that a remote operator can "take responsibility for several self-driving Pods, monitoring them when in autonomous mode and taking active control of a vehicle for unforeseen or more complicated maneuvers, such as parking at a loading dock." [5].

Altogether, these examples suggest that remote operators are anticipated to monitor or assess operation of automated vehicles, to provide assistance or guidance to automated vehicles in tricky situations, or even to actively control automated vehicles in situations that the automated driving system cannot resolve on its own. That is, remote operation is likely to be applied at strategic, tactical and operational levels of control [6]. At the same time, a recent study by Scania and RISE Research Institutes of Sweden concluded that there are still many unaddressed challenges when it comes to human factors in remote operation [7]. Examples of these challenges include:

- What are differences between operational, tactical and strategical control levels from a human factors perspective when it comes to road automation?
- What is the maximum number of vehicles that can be operated by one human operator simultaneously?
- How do we define Operational Design Domain (ODD) for remote operation?
- What are the methodologies and tools required for describing properties of and prerequisites for different ODDs?
- What education and certification is needed for remote operators?
- How will certification function with regard to regulatory aspects and frequent updates of automated driving systems?
- How could classical human-centric automation issues like trust, responsibility, automation surprises, boredom and vigilance be mitigated?
- How should a remote HMI be designed to accommodate different remote operation roles and switching between these roles?
- How can HMI be designed to enable people physically present in a traffic situation to support a remote operator?
- What are the tasks of remote operators for different ODDs?
- How much training is required for each task, and how do we train operators to handle edge cases?

While all these challenges are equally important and urgent to address, Scania and RISE are in their ongoing project Heavy Automated Vehicle Operator Center - Requirements and HMI (HAVOC) [8] investigating how remote operation centers should be designed to allow a remote operator to engage and switch between different remote operating roles, as well as to operate multiple vehicles simultaneously. The project is co-financed by the Strategic Vehicle Research and Innovation (FFI) program and is expected to be finalized in the beginning of 2021.

2.3 Highlights from Other Presentations

2.3.1 Panel Session A: Setting the Scene

This first panel session was aimed at introducing the goals of the session and provide an overview of remote support for automated vehicle operations with different examples of concepts of operations and the solutions developed. Four speakers have delivered a presentation followed by a panel discussion with room for Q&A. In addition to the work presented by Olav Madland (Sect. 2.1) and Azra Habibovic (Sect. 2.2), Elliot Katz, Co-founder and Chief Business Development officer at Phantom Auto (U.S.) and Elisa Bin, Research Engineer & Facilities Coordinator, KTH Royal Institute of

Technology (Sweden) gave a presentation. Katz discussed the remote control from his own experience at Phantom auto with his presentation: *Remote Operation: The Key Ingredient for Deploying Autonomy.*

The Phantom Auto Solution is described by Katz as way to enable humans to safely remotely monitor, assist, and drive any unmanned vehicle from thousands of miles away, with a software suite that is compatible with:

- All types of vehicles
- All wireless networks
- All hardware platforms

Its highly-integrable software allows enterprises to remotely operate all types of unmanned vehicle and robot fleets using commodity hardware (Fig. 3).

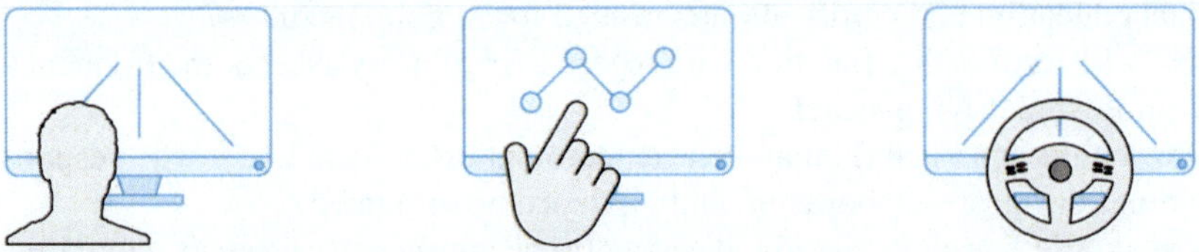

Fig. 3. Schematic overview of Phantom Auto Solutions: Remote Monitoring, Remote Assistance and Remote Driving, retrieved from presentation slides Elliot Katz.

The software powers unmanned operations, such as: Forklifts, Robots, Trucks and Cars. A use case was presented regarding forklifts. Customers are using Phantom's software to deploy unmanned forklifts in multiple ways:

1. Enabling autonomous forklifts to be deployed and scaled
    a. Remotely monitor and assist multiple AGVs/AMRs at once
    b. Increase vehicle uptime and efficiency
    c. Expand the vehicle's operational capabilities
2. Fully remotely-operated forklifts
    a. Reduce safety/health risks for warehouse employees
    b. Expand and diversify labor pool
    c. Dynamically distribute labor across vehicles and sites (e.g. on-call "pinch hitters" can instantly "teleport" anywhere)

GEODIS, one of the largest 3PLs in the world with 165,000 customers in 120 countries, is using Phantom's software to enable their employees ("digital drivers") to remotely operate forklifts from up to thousands of miles away.

Key Benefits:

- Increased operational health & safety
- Expanding the labour pool to underrepresented groups
- Increased operational efficiency and productivity, enabling Geodis to "teleport" drivers when and where they are needed the most

Finally, Katz presented three key takeaways and states that remote operation transforms material handling by:

1. Increasing operational **health and safety**
2. Increasing **labour accessibility**
3. Increasing **productivity and operational resilience**

The next presenter Elisa Bin working for the KTH Royal Institute of Technology in Sweden talked about *Automated Vehicle Traffic Control Tower (AVTCT), towards safe and connected transport system.* The automated vehicle traffic control tower (AVTCT) centralizes the decision and can act as an economic and safe backup of automated systems. It can also improve efficiency by improving fleet management and traffic flow. In AVTCT, one person can manage multiple automated vehicles, take actions upon request, and take over the control after system failures. One role of AVTCT is assuring the traffic safety and increase traffic efficiency. Another role could be coordinating among the fleets, infrastructures, service providers and traditional road users. AVTCT can act as a decision maker and can also be a decision support system for automated vehicles in dynamic driving scenarios. The control tower concept has been widely applied in aviation, marine and railway. However, the context is different for automated on-road driving and automation in the air, on railway and in the water. The AVs need road network and get more complicated interactions with surroundings infrastructure. The scale that AVs cover in transport is also broader and more complex than those aforementioned transportation modes.

The AVTCT is based on the concept of connected control towers. Figure 4 shows how a road traffic control tower, a fleet owner control tower and a confined area control tower (e.g. port or airport) can interact. Such an interaction and exchange of data may strongly improve the situational awareness of the AV fleet operator and thereby enable AV guidance. Relevant situational awareness data includes: maps, traffic flow, road works, traffic hazards, road conditions, weather, roadside equipment data, mobile coverage information, emergency vehicle approach, etc. This data is then combined with vehicle data such as position data, vehicle status information, vehicle events, features detected by the vehicle sensors, etc.

The primary tasks of the AVTCT when in operation is to supervise the vehicles and act upon the situational awareness if necessary or requested by the vehicles. Action support includes dynamic replanning, setting the AV mode (on, off, reduced speed), time/distance assessment to incidents, re-routing and if needed: vehicle control by activating pre-configured commands or tele-operation by driving the vehicle remotely.

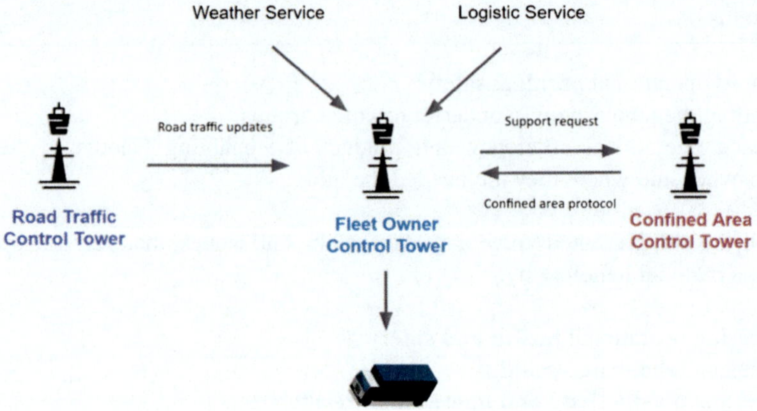

Fig. 4. Illustration of the automated vehicle traffic control tower (AVTCT), retrieved from presentation slides Elisa Bin

2.3.2 Panel Session B: Technological Perspective

Panel session B focused on features and technological aspects related to remote support, such as: connectivity and communication, vehicle and roadside surveillance equipment, standardization, scalability, cybersecurity, etc. Three speakers have delivered a presentation followed by a panel discussion with room for Q&A.

Firstly, Kiel Clasing, Manager Business Development and Timothy Gallagher, Senior Account Manager presented their work at Oceaneering (U.S.) presented: *Carry-over opportunities of remote command and control technology used in other industries.*

During their presentation Clasing and Gallagher presented relevant technologies within the Oceaneering industry and its link to remote support for autonomous vehicles. They identified the following enablers regarding remote support:

- Reduce dependency for on-board safety operators
- Facilitate L4 certification in mixed traffic environments
- Leverage data/comms from road, signals and other infrastructure
- Interact with passengers & other road users remotely
- Increase supervisory & monitoring capabilities

There are many other enablers one could think of. Via an interactive poll with the audience it was found that 'Increase adoption/readiness of AV's' is one of those enablers which seemed highly relevant for the Breakout session audience.

Lastly, Clasing and Gallagher highlighted several relevant use cases and potential applications in relation to Oceaneering. Figure 5 shows these use cases/applications.

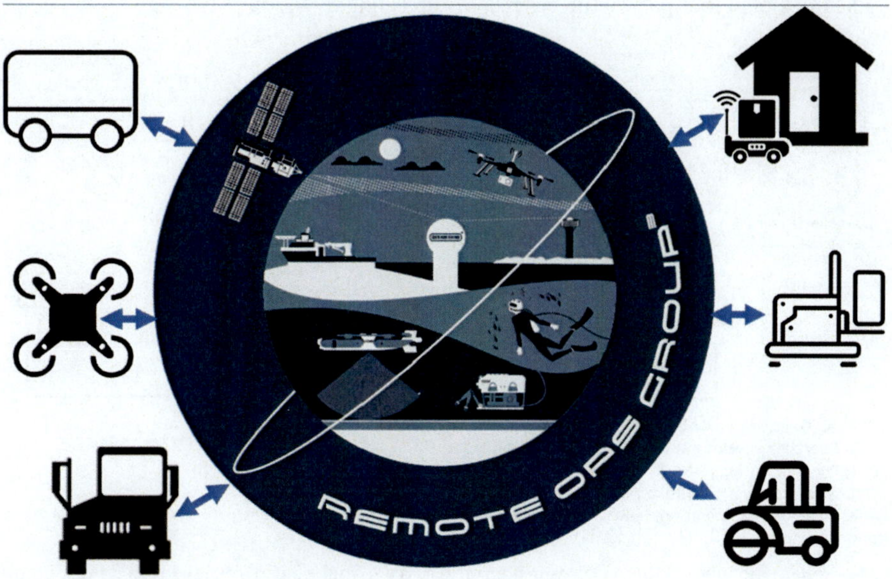

Fig. 5. Overview of potential applications and links to Oceaneering, retrieved from presentation slides Kiel Clasing/Timothy Gallagher

- Connected/Automated Vehicles
- Personal Delivery Devices
- UAS/Drone
- Long Haul Logistics
- Trailer Yard Management
- Freight/Material Handling
- Road/Lawn Maintenance

Secondly, Dr. Edward Griffor, Associate Director, National Institute of Standards and Technology (U.S), presented on *Measuring the Safety of Automated Driving Systems (ADS).*

Griffor starts with a quick overview of the concepts regarding AVs such as SAE levels, ADS and the Operating Envelope Specification (OES), which is also known as Operational Design Domain (ODD). He defines the OES as a structured description of the operating environment for driving, suitable to support formal reasoning about that environment in testing and certification applications and in real-time driving conditions. An instance of an OES comprises the dimensions of the operational state space (whether chosen by the manufacturer, developed from a relevant scenario set, or defined de novo) sufficient to enable reasoning about the state space.

By means of an ADS logic chart, the process of an AV trip is explained (Fig. 6):

1. Receives a 'Trip Goal';
2. Provides 'Path Plans' based on HD Map and time and present position;
3. Monitors the status of path plan development and checks and execution;
4. Executes plan that passes the criteria of the Logic.

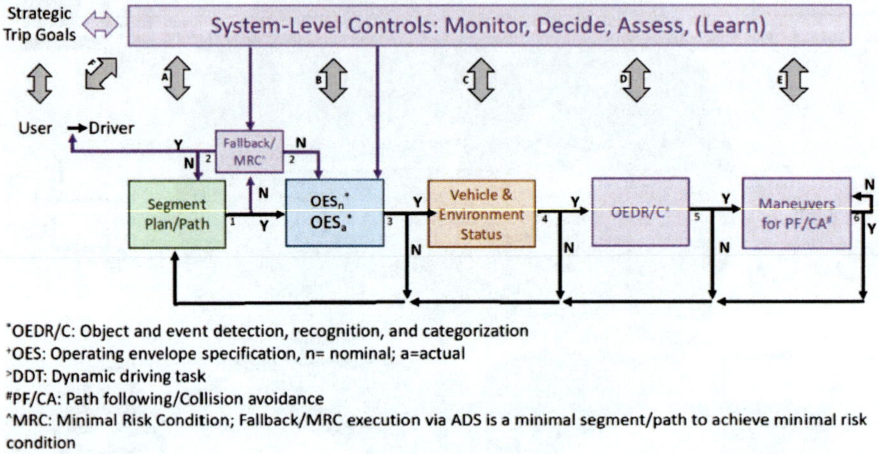

Fig. 6. Schematic view of the ADS logic chart for an AV to make a trip, retrieved from presentation slides Dr. Edward Griffor.

Lastly, Griffor presents a simulation demo for an AV: Drive Cycle and Events (Fig. 7). This simulation entails varying speed limits, AEB (Advanced Emergency Braking), ACC (Adaptive Cruise Control) and Events (OEDR).

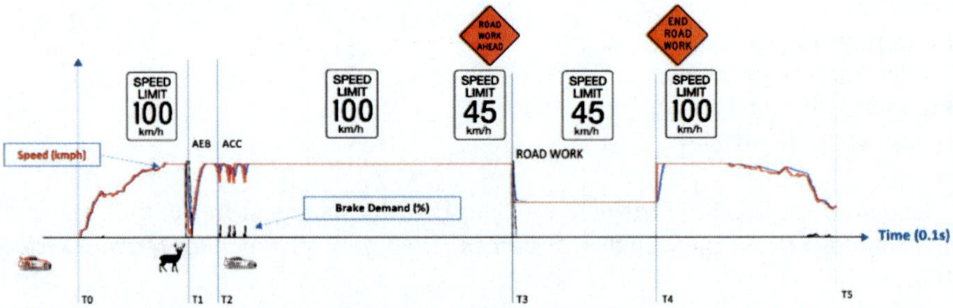

Fig. 7. Drive Cycle and Events Demo, retrieved from presentation slides Edward Griffor.

For a given vehicle with certain specifications (weight, aerodynamics, tire rolling resistance) the safety assessment is based on:

1. Whether the simulated vehicle drives at the designated speed
2. Whether the simulated vehicle adapts its speed and perdorms enough braking to safely navigate the following events:

- obstacle detection (maintain safe distance between vehicle and object)
- car following (adaptive cruise control should not impact comfort adversely, i.e., multiple speed changed in a short time interval)
- road work/construction (vehicle correctly interprets the beginning and the end of roadway events, e.g., speed limit changes)

To conclude Griffor states that Operating Envelope Specification (OES) and co-simulation to assess the safety of automated driving systems.

The panel session was closed by a third presentation from Andrew Phillips, Manager, Connected and Automated Vehicle Safety Programs (Canada) on *Road safety considerations for the development and testing of remote support technologies.*

To start of Phillips emphasizes that road safety is a shared responsibility in Canada and that each of the areas, i.e. federal, provincial/territorial and municipal take their part in this responsibility. To support the safe testing and deployment of connected and automated vehicles the safety regimes require adaptation. Over the past years, several documents and guidelines where published to secure this safety aspect: Amendments to the Motor Vehicle Safety Act (March 2018), Testing Guidelines (June 2018), Transport Canada Safety Framework (February 2019), Safety Assessment for Automated Driving Systems (February 2019), Vehicle Cycle Security Guidance (May 2020).

Currently additional safety best practices are examined which relate to the following themes:

- Authorization procedures (e.g. authorization checklist);
- Assessing test vehicle safety (safety assessments);
- Developing safety management plans for trial operations;
- Safety drivers and safety driver training;
- Passenger safety;
- Engagement with first responders and law enforcement;
- Low speed shuttle safety;
- Safety of remote support applications.

Griffor furthermore discusses the potential benefits concerning remote support:

– Remote dispatcher and remote monitor applications may help to enhance:

  o The efficiency of the automated vehicle's operations;
  o Passenger safety and security;
  o Information and service provision to passengers.

Remote assistance and remote driving applications in turn may enhance safety by helping to overcome current limitations of ADS technologies as they continue to be refined and developed. This could include assistance when the ADS:

- Exits its operational designed domain (due to a change in weather for example);
- Encounters a rare or particularly complex scenario it has not been designed to navigate (e.g. edge case/corner case).

Griffor points out that remote support may also pose unique safety challenges that could also require careful consideration. Figure 8 shows a non-exhaustive overview of some potential safety challenges associated with remote support applications. It must be noted that safety risks may vary significantly depending on the remote application in question, and the complexity of the driving environment (e.g. controlled, low speed environment versus mixed traffic at highway speeds).

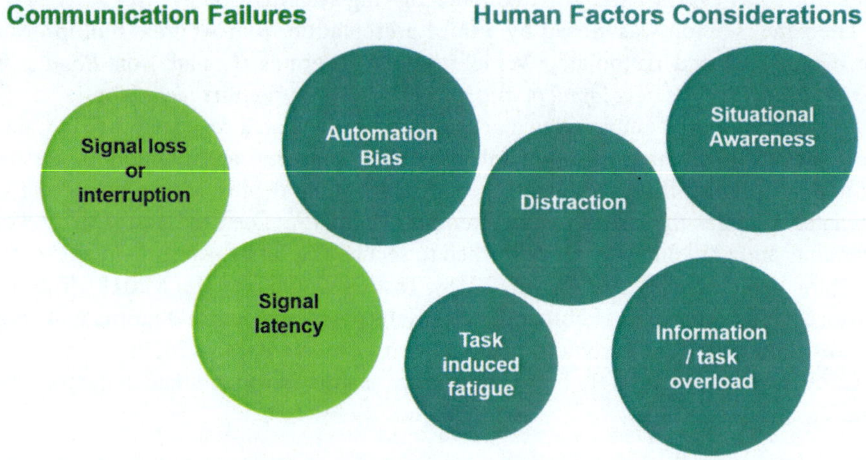

Fig. 8. Overview of some potential saftey challenges associated with remote support applications, retrieved from presentation slides Andre Phillips.

Finally, Griffor closed with the following key takeaways:

- Safety implications and best practices could benefit from further research and analysis by the international community, particularly for remote driving where risks are likely greatest
- A graduated approach to testing – where complexity and risk are incorporated in a gradual, iterative fashion, can help to support the safe testing and adoption of remote support technologies

Dialogue between industry, the public, and authorizing jurisdictions will also be important to foster. For example, industry should consider elaborating on remote support applications and their safety validation efforts as part of voluntary safety assessment publications.

3 Conclusions and Next Steps

The key findings and lessons learned from this session on remote support for automated vehicle operations are:

- A proven success factor for vehicle operation without a safety driver on board and use of a remote control room has been a stepwise approach. This allowed all stakeholders and especially the regulating authorities to incrementally built experience and trust with such operations;
- There are many questions to answer with regards to human factors in remote operation. New risks of failure arise as well as new challenges mostly linked to the limitations of what a (human) operator can realistically handle;
- The capabilities and technologies are there, it is mainly a matter of going into details on definitions and standards regarding for instance ODDs.

The implementation needs and topics for future research which were identified during the closing panel discussion are:

- It is important to clearly define the role of the human operator. Some tasks might be more of a supervision rather than controlling nature;
- A big aspect is the ODD and related degree of intervention, which depends on many factors. For example, intervention of an operator on a defined route with no impeding traffic compared to an open road and mixed traffic;
- Research is needed into higher speeds and resulting increased safety risks;
- Further studies should investigate edge cases, i.e. situations in which a remote operators handles one situation and then suddenly must switch to a completely different one;
- Remote support is only one element within a much large domain. It is needed to look at for instance ODD from a system-to-system perspective beyond the scope of only the vehicle.

References

1. Argo, A.I.: Safety report: developing a self-driving system you can trust (2021). https://www.argo.ai/wp-content/uploads/2021/04/ArgoSafetyReport.pdf. Accessed 25 Sept 2021
2. Beggin, R., Hall, K.: Remote control for self-driving cars isn't required anywhere. Why? It's complicated. The Detroit News (2021). https://www.detroitnews.com/story/business/autos/2021/04/06/remote-control-driverless-cars-not-required-why-its-complicated/6840041002/. Accessed 25 Sept 2021
3. Ackerman, E.: What full autonomy means for the Waymo driver. IEEE Spectrum (2021). https://spectrum.ieee.org/full-autonomy-waymo-driver. Accessed 25 Sept 2021
4. Nuro: Delivering Safety. https://nuro.sfo3.digitaloceanspaces.com/Nuro-VSSA-2021_Final.pdf. Accessed 25 Sept 2021
5. Einride: Einride showcases one operator, multiple vehicle capability at a customer site (2020). https://www.einride.tech/press/einride-showcases-one-operator-multiple-vehicle-customer-site. Accessed 25 Sept 2021
6. Michon, J.A.: A critical view of driver behavior models: what do we know, what should we do?" In: Human Behavior and Traffic Safety, pp. 485–524. Springer, Boston (1985). https://doi.org/10.1007/978-1-4613-2173-6_19

7. Habibovic, A., Andersson, J., Castor, M., Meiby, L., Rizgary, D.: Human factors of remote control. SAFER Prestudy report (2020). https://www.saferresearch.com/library/final-report-human-factors-related-remote-control-automated-heavy. Accessed 25 Sept 2021
8. VINNOVA: Heavy Automated Vehicle Operation Center (HAVOC) - Requirements and HMI design (2020). https://www.vinnova.se/en/p/heavy-automated-vehicle-operation-center-havoc---requirements-and-hmi-design/. Accessed 25 Sept 2021

Part V: Transport System Planning

Ensuring Strong Public Support for Automation in the Planning Process: From Engagement to Co-creation

Leah Kaplan[1] (✉), Siegfried Rupprecht[2], Monica Grosso[3], Nikolas Thomopoulos[4], Wolfgang Backhaus[2], Maria Alonso Raposo[3], Biagio Ciuffo[3], and Daniel Franco[2]

[1] Department of Engineering Management and Systems Engineering, The George Washington University, 800 22nd St NW, Washington, DC, USA
leahkaplan@gwu.edu

[2] Rupprecht Consult, Clever Strasse 13-15, Cologne, Germany
{s.rupprecht,w.backhaus,d.franco}@rupprecht-consult.eu

[3] DG Joint Research Centre, European Commission, Via E. Fermi 2749, 21027 Ispra, VA, Italy
{monica.GROSSO,RAPOSO,Biagio.CIUFFO}@ec.europa.eu

[4] University of Surrey & WISE-ACT Chair, Stag Hill, University Campus, Guildford, UK
n.thomopoulos@surrey.ac.uk

Abstract. Hailed as the next transportation revolution, automated vehicles (AVs) are expected to have dramatic impacts on the environment, economies, and society. Potential benefits of, as well as any potential burdens from, AV deployment will depend in large part on whether and how individuals will use them. Gaining a better understanding of public attitudes towards AVs can thus provide important insights into the future of automated transportation. This conference session brought together participants from the academic, planning, and commercial business development sectors to learn about the status of public attitudes towards and preferences for AVs, discuss transferable results from prior engagement projects and promotion campaigns, exchange strategies for engaging citizens in the AV planning process, and coordinate strategies at an international level. During the session, the presenters highlighted key findings and lessons learned from prior research and engagement efforts, as well as outlined areas for future work and research.

Keywords: Autonomous vehicles · Automated vehicles · Co-creation · Public engagement · Citizen engagement

1 Introduction

Hailed as the next transportation revolution, automated vehicles (AVs) are expected to have dramatic impacts on the environment, economies, and society. In fully-automated vehicles, defined as Level 5 AVs by SAE International, a computer system would operate the vehicle at all times and in all conditions [1]. Expected benefits from AVs are even

G. Meyer and S. Beiker (Eds.): ARTSymposium 2021, LNMOB, pp. 167–183, 2023.
https://doi.org/10.1007/978-3-031-11112-9_13

more significant for vehicles that are *connected* and *automated* (CAVs), meaning that the vehicles could communicate with each other and with the surrounding infrastructure. Both AVs and CAVs could play crucial roles in reducing greenhouse gas emissions from the transportation sector through efficiency gains, reduction in vehicle ownership, and the potential for AVs and CAVs to be electric [2]. Given that the transportation sector is one of the largest contributors to emissions, emissions reductions from AVs could play an important role in meeting sustainability goals forwarded by the Intergovernmental Panel on Climate Change, the 2015 Paris Climate Agreement, the European Green Deal, and country-level climate goals [3–5].

Yet the potential benefits of AVs are far from certain [6, 7]. Potential benefits of, as well as any potential burdens from, AV deployment will depend in large part on whether and how individuals will use them. Therefore, it is pertinent to ensure that the user perspective is an integral part of AV deployment and relevant ICT infrastructure planning [8]. Gaining a better understanding of public attitudes towards AVs can thus provide important insights into the future of automated transportation. These insights can allow transportation planners to better plan and prepare for an automated future. Beyond simply planning for an AV future, engaging the public in consultation and co-decision processes can help to shape transportation futures in a manner that aligns with public values through user-centered solutions. By engaging with the public as part of the planning process, researchers and transportation planners can better direct AV development and deployment toward desired environmental, economic, and social outcomes whilst meeting wider policy targets at the same time.

The relevant ARTS21-TRB session which provided the input for this chapter focused on sharing key outputs from ongoing engagement efforts, as well as best practices for conducting public engagement activities with an AV focus. Session presenters had worked on a range of projects spanning the globe, including a wide range of projects and activities in Europe (CoEXiST, the WISE-ACT[1] survey and focus groups in collaboration with the European Commission Joint Research Centre, H2020 SHOW, Urbanism Next Europe), the United States (TOM-NET, Our Driverless Futures), and Asia (BOLDLY). Given the session's focus on public engagement as part of the planning process, it also included panel discussions with individuals from the policymaking domain. These policymakers included representatives from the European Commission and the United States Department of Transportation. This approach of including practitioners, researchers, policymakers (e.g. projects), and user/wider societal input (e.g. user surveys) embodies the Quadruple Helix approach [9] in practice on this topic of emerging interest across continents.

Overall, the objectives for the session were for participants from the academic, planning, and commercial business development sectors to:

- Learn about the status of public attitudes towards and preferences for AVs, including differences among various social and demographic groups (e.g., gender, age, ethnicity, individuals with disabilities) and across various geographies (e.g., urban/rural, multiple countries);

[1] https://www.wise-act.eu.

- Discuss (transferable) results from prior engagement projects and promotion campaigns;
- Exchange strategies for engaging citizens in the AV planning process; and
- Coordinate strategies at an international level given that sustainability is an international challenge.

The remainder of this chapter will share information from the conference session, as well as expand on the session's presentations and discussions to provide a more comprehensive portrait of public attitudes towards AVs and effective engagement approaches. Section 2 provides a summary of key themes from the session discussion: 1) research on public perceptions of AVs across various sociodemographic groups and geographies, 2) transferrable results from prior engagement projects and promotion campaigns, and 3) strategies for engaging the public in AV planning and deployment. Finally, the conclusion offers reflections on key takeaways from the session, and provides suggestions regarding how the discussion might continue and expand in the future. The conference session on which this chapter is based was supported by the WISE-ACT COST Action 16222.

2 Summary of the Discussion

The conference session covered a number of different topics pertaining to public knowledge of and interest in automated transportation technologies. The following sections summarize the current status of public support for automated transportation technologies in different social groups and geographies, describe transferrable results from prior engagement and promotion campaigns, and offer strategies for public engagement and involvement in the planning and implementation of regular, automated services.

2.1 Status of Public Support in Different Social Groups and Geographies

Effective strategies for garnering public support for automation necessitate an understanding of current public perceptions of, hopes for, and reservations about automated transportation modes. Individuals have unique transportation needs and routines and their attitudes towards automation are similarly varied. Policymakers from Europe, Japan and the United States stressed that strong public support is essential to increase acceptance, where applicable, and deploy automated transport services successfully. A growing body of literature has investigated AV attitudes amongst different demographic groups. Further, additional studies have surveyed attitudes towards AVs in different geographies to determine how varying transportation landscapes might also influence public attitudes. This section aims to provide an overview of research findings regarding public awareness of automated vehicle technologies; public attitudes towards AVs amongst different demographic groups; insights into additional non-demographic factors that influence AV attitudes; and how support for AVs varies by spatial (e.g. urban, rural) or social features (e.g. gender, culture). We limit our discussion to research regarding automated vehicle use for personal transport. Most of the referenced studies defined automated vehicles in terms of fully-automated (SAE level 5) vehicles, though some studies asked questions more generally about "automated" vehicles, without specifying a level of automation.

2.1.1 Awareness of AV Technologies

A starting point for increasing public support for AVs is determining whether the public is even familiar with AV technologies. Though AVs have yet to reach large-scale deployment, awareness of AVs is growing, with some studies finding high levels of at least mild awareness of AVs. In one survey of four U.S. cities, 49% of respondents reported being at least somewhat familiar with AVs and only 15% of respondents reported never having heard of AVs [10]. The study also found that awareness of AVs increased with higher education; higher reported enjoyment of trying new things; increased familiarity and use of ride-hailing; and higher income [10]. Another U.S. study found even higher levels of awareness. Approximately 97% of an over 5,000 person sample taken from across the U.S. reported having heard at least "a little" about AVs [11]. These high levels of awareness may in part stem from the various pilot tests taking place across the United States.

Levels of AV awareness are also relatively high in other countries. Moody et al. [12] surveyed individuals across 51 countries and found that 55.9% of people were "a bit" aware and 19.4% were "very aware" of AVs—defined by having seen, heard, or read about driverless cars. In other multi-country surveys in Europe, over 60% of respondents said they had heard, read, or seen information about AVs in the last 12 months [13, 14]. All of these studies offer promising signals that ongoing communication and outreach efforts may be succeeding in increasing AV awareness at an international level. Nevertheless, policymakers must recognize that due to the nature of reporting findings through such global surveys, high global awareness proportions in aggregate level may mask low levels of awareness in some countries, particularly within under-represented socio-economic groups. Indeed the WISE-ACT survey distributed in 25 COST[2] countries across Europe found that public awareness of AVs varies significantly by country [15]. More precisely, the public appears to have heard more about AVs in northern and western Europe (e.g. Germany, Iceland, UK) compared to southern and eastern Europe (e.g. Bulgaria, Romania). This is not surprising given the spatial distribution of AV trials across the continent [16].

2.1.2 Who Will Adopt AVs?: Demographics and Attitudes of AV Adopters

The question of who will adopt AV technologies remains at the forefront of many AV developers' and transportation planners' minds. To identify this question of "who", researchers have examined AV attitudes amongst various demographic groups and studies summarizing such findings have been emerging [15, 17, 18]. A contemporary review of stated preference surveys and choice studies on AVs, Gkartzonikas and Gkritza [19] observed that the majority of studies targeted the general population as the study sample. With representative samples of the respective study region, researchers then looked for preference patterns based on various demographic and non-demographic characteristics.

Within the existing literature, general agreement exists that men are more likely to use AVs than women, with some studies attributing this gender difference to personal security concerns for women [10, 11, 15] as has also been reported previously about

[2] COST (Cooperation in Science and Technology) is the longest running funding scheme in Europe: www.cost.eu.

non-automated shuttle services in e.g. Mexico City [20]. Other groups that appear to hold more positive views about AVs or who express the greatest willingness to use or buy AVs are individuals who are younger, highly educated, higher income, and fully employed [10, 11, 21, 22]. Despite the widespread view that age, education and disability have been found as significant attributes, it has proven difficult to identify global trends across demographic characteristics. Contradictory findings exist, for instance, regarding linkages between AV support and income [11, 14, 22, 23].

Non-demographic characteristics also appear to influence attitudes towards AVs. The latter has been particularly evident through surveys across Europe (e.g. Eurobarometer, WISE-ACT). Individuals who are more familiar with ADAS (Advanced Driving Assistance Systems), ACC (Automated Cruise Control) or broadly more tech-savvy—characterized in some studies by current use of smartphones, text messaging, Facebook, and transportation apps—favor AV use over non-automated modes in some scenarios and express higher levels of AV interest [11, 14, 24].

Individuals' current mobility routines also shape their views towards AVs. Kassens-Noor et al. [25] found that individuals who regularly use non-automated transit modes expressed less interest in using automated transit. Although automated transit may not appeal to current transit users, Dong et al. [26] found that current infrequent or non-users of public buses were actually open to riding in an automated bus, even more so than frequent riders. Car owners and users tend to have both a higher awareness of AVs and more optimistic views of current and future AV safety [12]. Though such findings provide some hope of luring current private vehicle drivers onto automated transit modes, other studies seem to signal that individuals may stick with the modes as before, but just shift to automated versions. For example, Polydoropoulou et al. [15] found that current car use positively affected choice of private AVs and current public transport use positively affected choice of shared AVs in the majority of countries surveyed: Cyprus, Finland, Greece, Hungary, Iceland, Israel, United Kingdom. Interestingly there is wide variation regarding willingness to share journeys on an automated vehicle depending on the number and gender of co-passengers i.e. ride-sharing (Fig. 1).

In comparing the weights of demographic and non-demographic factors, attitude classes (e.g., pro-technology, low concern for data and privacy, interest in driving) and current mobility profiles were associated with greater impacts on AV attitudes than demographic characteristics [27, 28]. These findings underscore a potential need for future studies to aim for not only demographic representativeness but diversity in current mobility routines amongst survey respondents.

Overall, many demographic and attitudinal characteristics influence individuals' attitudes towards AVs [19, 29]. While demographic trends can help guide the design of public communication strategies, AV developers and policymakers must recognize the multifarious factors that influence individual decision-making around AVs and be prepared to address a number of public concerns to increase acceptance and eventually uptake.

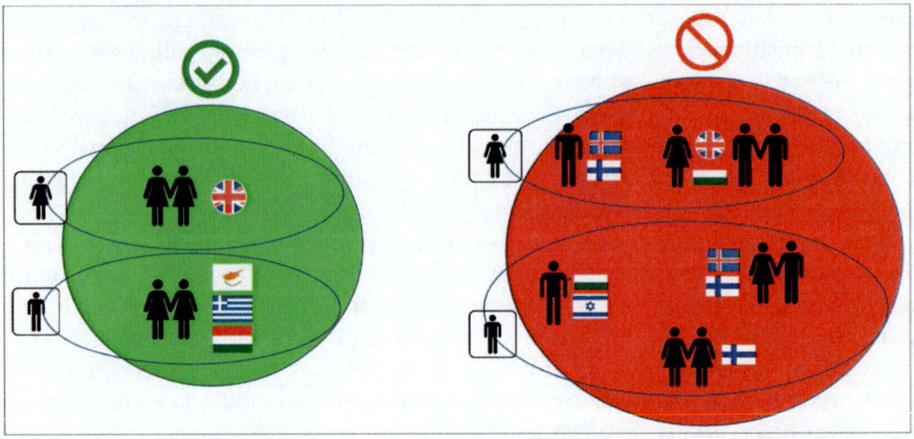

Fig. 1. Willingness to share an AV in seven countries, according to the number of co-passengers and their gender. The green circle on the left indicates the conditions under which respondents from different countries (indicated by the flag icons) were willing to share an AV. The red circle on the right indicates the conditions under which respondents from different countries were not willing to share an AV. Responses are broken down by gender (responses from women on top and responses from men on bottom). Example interpretation: Women from the UK were willing to share an AV with two women but not with one woman or with two men. (Figure from Polydoropoulou et al., 2021).

2.1.3 Geographic Trends in AV Adoption

The unique transportation landscapes in different countries and regions appear to yield similarly varying public attitudes towards AVs. Studies that surveyed members of the public across multiple countries found inter-country variation [12–15, 21, 30]. Moody et al. [12]'s comparison study of 51 countries identified a small but significant country-to-country variance in perception of AV safety, even after controlling for sociodemographic characteristics. Cross-national studies have also probed—and found differences in—preferences for different automated transportation modes. Amongst the 721 individuals from Israel, the United States, and Canada surveyed by Haboucha et al. [30], 54% of North Americans preferred non-autonomous cars over private autonomous cars or ridehailing services; Israeli preferences were split relatively evenly between the three potential modes (35% prefer non-autonomous cars, 36% prefer private AVs, and 29% prefer ridehailing).

As part of the 2019 project "Our Driverless Futures", 24 cities across Europe, North America, and Asia hosted day-long deliberations on the future of automated mobility. All host sites followed the same deliberation protocol and data collection procedures, allowing for results comparison across sites. Outputs from the deliberations revealed notable differences between different countries, as well as some similarities across sites. Participants in U.S. cities, for instance, expressed higher levels of support for an individual ownership model of AVs than participants from the Austrian deliberation sites [31, 32]. Outputs from the global project and additional research findings reveal not only inter-country variation, but also intra-country variations in public AV preferences [10,

31, 33]. Equally, the WISE-ACT 2020 survey and the JRC focus-groups held online during 2020 because of COVID-19 restrictions, highlighted similar concerns through their findings.

One often discussed geographic divide is the difference between rural, suburban, and urban areas. In quantitative studies, urban dwellers have demonstrated greater openness toward AV use than their suburban and rural counterparts [19, 24]. Indeed many companies are targeting their deployment efforts toward urban areas as highlighted by the large number of urban AV trials [16]. Nevertheless, rural areas whose lack of density makes efficient public transportation difficult also stand to benefit from automated technologies. In a series of focus groups held in both an urban area and a rural area (Baltimore, Maryland and Cumberland, Maryland) in the United States, rural residents actually expressed greater excitement about AVs but also concerns that the technologies would never reach their area [33]. Similar findings were reported in the 2018–19 Sciencewise[3] deliberation event across five UK cities.

Just as policymakers and transportation planners should work to expand pilot testing to diverse geographic areas, researchers should similarly expand research on public preferences. Studies that focus only on specific cities or countries—often those in which AV testing is occurring—perpetuate perceptions of limited public support in other cities and countries [16]. Thus it is recommended (e.g. WISE-ACT) to expand both the spatial and socio-economic dimension of AV trials to include more diverse user groups and engage with interested citizens more directly.

2.2 Transferrable Results from Prior Engagement and Promotion Campaigns

This ARTS21-TRB conference session included presentations on prior AV engagement and promotion campaigns that have been held in cities around the world. The projects discussed ranged from more traditional promotion events to deliberative fora and focus groups with both transportation experts and members of the public. In Japan, the company BOLDLY conducted AV promotion campaigns in the town of Sakai and the Haneda Innovation City. In Summer 2019, Paris-based public consultation organization Missions Publiques organized the public deliberation project "Our Driverless Futures" in cities across Europe, the United States, and Asia. The European Commission's Joint Research Centre (JRC) also carried out a series of focus group discussions with both experts (e.g. WISE-ACT) and non-experts in transportation in Europe. We describe here these three projects, linking them with the international surveys and research activities previously mentioned (Sect. 2.1).

2.2.1 BOLDLY

BOLDLY's AV promotion campaigns in the Japanese towns of Sakai and Haneda Innovation City were centered around the goal of commercializing an autonomous bus and "aim[ing] to create a town where residents of all generations can move safely and conveniently" [34]. Launched in November 2020, the initial route for the bus in Sakai covered a 5 km round trip that provided service to important facilities around the town including

[3] https://sciencewise.org.uk/projects/connected-and-autonomous-vehicles.

the bank, hospital, and the local elementary school. After an initial pilot phase, BOLDLY expanded the bus route to over 20 km in July 2021. This phased approach allowed residents to become familiar with the bus prior to expanding its service. It also allowed the company to troubleshoot problems on a smaller scale. For instance, the bus was initially causing traffic jams due to its stops. Partnering with residents who allowed the use of part of their property as a bus stop ultimately resolved the traffic problem. Another key learning from the Sakai project was the value of social media promotion of the bus. BOLDLY found that an effective recruitment strategy for new passengers was sharing residents' experiences through social media [34].

2.2.2 Our Driverless Futures

The *Our Driverless Futures* project brought together more than 2,500 citizens across 24 cities for day-long deliberations about automated mobility. The deliberations consisted of 5 discussion sessions that were the same across all of the cities, and one "local session" that each city crafted in collaboration with local stakeholders. This project structure allowed for comparisons of citizen views across different countries, while still allowing for deeper discussion on city-specific issues. An important learning from the project was that building such a widespread partner coalition takes time. For instance the project lead within the United States, the ASU Consortium for Science, Policy & Outcomes, required almost a year to identify and train cities for the four U.S. deliberations [35].

2.2.3 The European Commission Joint Research Center (JRC) Focus Groups

The JRC, together with the German Aerospace Centre (DLR) and the University of Cantabria, carried out 15 focus group discussions on the topic of fully automated and connected vehicles with experts including the WISE-ACT community and non-experts in transport. The discussions took place from June 2020 to January 2021 involving a total of 72 participants, 40 with and 32 without expertise in transport, with an average age of 41.2 years. During the discussion, participants shared their prior knowledge and experiences with advanced driver assistance systems (ADAS) and automated driving; the threats and benefits they associate with AVs; their current mobility needs; and their overall perceptions of CAVs. The main results of the discussions informed researchers about the importance of developing trust around CAVs in relation to different topics including individuals' willingness to use CAVs, challenges with mixed automation levels, and attitudes towards sharing public space with CAVs.

When analysing the answers provided by transport experts and non-experts, researchers found clear agreement regarding main points discussed and both expert and non-expert views and opinions on the topics discussed. The main difference was that expert participants supported their views with more elaborated speculations—concepts and arguments deriving, most likely, from their professional activities and background (e.g. within WISE-ACT). Besides perceived potential benefits—including safety, accessibility and travel efficiency—participants mentioned concerns regarding safety, legal responsibility, and privacy. Participants also expressed the need to trust the technology and to have a guarantee that it is safe before using it. Some participants mentioned that as vulnerable road users, they expect the infrastructure to ensure their safety when they share the space with CAVs. All of the identified concerns, if not carefully considered by stakeholders, will negatively impact AVs acceptance by citizens and slow down their market deployment.

While participants expressed a high degree of willingness to use CAVs, many participants did not think about use in terms of their daily mobility routines. Instead, they considered CAV use for leisure trips (e.g., long-distance trips) reinforcing contemporary suggestions in the literature [18], or for instances when they would not have any other alternatives.

2.2.4 Key Takeaways from Prior Engagement Projects

Public "education" is Not Enough: Despite ongoing engagement efforts, members of the public still seem to generally prefer traditional cars over automated vehicles [21]. These preferences signal more than just a continued lack of knowledge of or understanding about AV technologies. In fact, existing research, such as the WISE-ACT survey which reported higher Willingness To Accept AVs in Cyprus and Slovenia compared to lower in Iceland and the UK [21], shows mixed results on whether experience with vehicle automation increase support for AVs. This finding about the role of education was also confirmed during the relevant international WISE-ACT workshop about Automated and Connected Transport education held in Riga in 2019. While some studies found that exposure to automation increases support for AVs [12, 14, 36, 37], other studies have found that increased awareness of AVs actually increased the probability that individuals would see AVs as dangerous or that they would report concerns with the technology [10, 38]. These results signal the impacts of AV promotion and education campaigns may be limited if underlying public concerns are not identified and addressed. As Suzanna Kraak—a policy officer at the European Commission—stressed during this ARTS21-TRB session, education is a two-way process which should form the backbone of effective citizen engagement and co-creation processes for the deployment of new mobility (e.g. AV) services.

Move Beyond the Safety Arguments: Potential safety benefits have been the main argument supporting AV deployment since the early days [39, 40]. AV developers must realize however, that individuals think about safety more broadly than traffic accidents. Some studies, for instance, have identified the potential importance of having an operator on board AVs to maintain social cohesion and take into account the needs of vulnerable users [14, 26]. Further, making the case for AVs must go beyond safety to address individuals' broader hopes and concerns. Identified hopes and concerns include sensitivities to travel time and reliability; desires to retain control; privacy and employment challenges; and concerns about sharing vehicles [10, 14, 15, 21].

Consider Benefits and Harms at Both the Individual and Societal Level: Prior engagement projects found that the public thinks about AV opportunities and threats at both the individual and societal levels [31, 32, 41]. To design effective engagement campaigns, planners and AV developers must address public hopes and concerns at both of these scales. Further, engagement efforts should focus not only on members of the public, but also on under-represented socio-economic groups and public authorities. One EU-based project found that some public authorities lack knowledge about AVs and do not have clear strategies towards effective AV planning and implementation, which raises specific concerns regarding sustainable urban mobility planning [42].

Start Planning Now But Be Prepared to Adapt: Although the evaluation of CAVs was positive among many focus group participants, it seems like the road towards public adoption of the technology will be a long process. Nevertheless, transportation planners should start preparing for implementation of CAVs in the transportation system and adjust preparations based on updated feedback from the public and stakeholders. The following section offers strategies for soliciting this type of feedback.

2.3 Strategies for Public Engagement and Involvement in the Planning and Implementation of Regular, Automated Services

A number of strategies for public engagement and involvement in the planning and implementation of AVs emerged from the session's presentations and discussions. This section details some of these general strategies.

2.3.1 Create Opportunities for Individuals to Experience AV Technologies

One successful strategy for increasing public understanding and awareness of AVs is to use computer models, simulators, and virtual reality demonstrations to allow individuals to "experience" AV technologies [42]. Moreover, in cities where pilot tests are taking place, it is valuable to create opportunities for members of the public and policymakers to interact with the technology.

Fig. 2. A local cake shop in Sakai, Japan modeled a cake design after the AV shuttle being tested locally. Image courtesy of BOLDLY.

During pilot testing in Japan, for instance, the company BOLDLY worked with local companies to create diverse, interactive engagement experiences that helped build community support, even amidst ongoing technical challenges [34, 43]. One creative partnership was with a local cake shop that started creating bus cakes to promote the AV shuttle being tested in the city (Fig. 2). Such an approach may facilitate citizen engagement and discussions at various levels.

2.3.2 Develop a Shared Vision and Communication Strategy

One of the outputs from European fora with policymakers was the importance of engaging with stakeholders and planners ahead of time to create a shared understanding of, and vision for, an AV future [42]. After developing a common strategy, planners should focus on using narratives to communicate that vision to the public. One narrative that proved valuable during BOLDLY's engagement projects in Japan was the idea of an AV bus as a "horizontal elevator"—a free service that connects individuals to critical services around the town. Another approach is to collaborate with professional storytellers. For the "Our Driverless Futures" deliberations hosted in the U.S., organizers worked with a documentary filmmaker to make briefing information about how AVs function informative, accessible, and engaging [31]. When crafting narratives about AVs, the session speakers agreed on the importance of framing AV discussions in the context of larger new mobility goals such as sustainability and equity [18, 32, 42].

2.3.3 Promote Co-creation and Living Labs

Co-creation approaches are becoming increasingly important for achieving successful integration of automated mobility services in urban planning processes. Citizens need to be able to understand the potential impacts and benefits from automated mobility. They need to be able to express their wishes, expectations, and concerns as part of being actively engaged in the co-development of automated mobility services. Engaging citizens in the co-design of future transportation systems would not only contribute to raising awareness about the added value of new technologies, but also ensure that the new technologies properly address people's diverse needs and expectations. Using open-framing dialogues and forums can create space for the public to bring up additional concerns that experts may not identify [33, 41]. Further, public forums, especially those with tailored formats to address different target groups, can add nuanced understandings to complement quantitative surveys [31, 32, 42].

Another effective approach to the co-creation process is the use of Living Labs. Living Labs can allow cities to move from mere testing of new mobility solutions to co-creation of such solutions. Living Labs are defined as *"user-centered, open innovation ecosystems based on a systematic user co-creation approach, integrating research and innovation processes in real life communities and settings"* [44]. Living Labs can benefit research and innovation (R&I) projects that aim to address societal challenges by putting citizens at the heart of the innovation development process. Living Labs apply a multi-stakeholder approach that usually follows the Quadruple Helix model—including stakeholders from the public sector, academia, citizens, and industry. Living Labs are able to actively engage users and public/private stakeholders in promoting the co-creation of value so as to benefit the economy, society, and the environment.

In the case of automated mobility, Living Labs could help citizens to better understand what automation would mean for their lives, set realistic expectations, remove fears/concerns, and build trust towards this new technology. Living Labs, along with citizen science hubs, could engage citizens and relevant private and public stakeholders starting in the early stages of development of automated mobility systems and services. This early engagement would allow for experimentation with automated mobility technologies and policies, ultimately helping to assess the potential implications of this new type of mobility. Living Labs could thus shape AV systems that support broad societal benefit and help cities achieve sustainability goals. In order to support transportation equity as a goal [45, 46], it is essential that engagement processes involves users from all societal groups, to ensure that all perspectives are heard and accounted for.

Finally, to leverage higher impact from global-scale initiatives, it is crucial to foster international cooperation among different players. This cooperation should focus on sharing knowledge and best practices to build up capacity for better coordination of R&I activities in the automated mobility field. International collaborations represent opportunities for scaling up AV research and testing activities, allowing for transferability and replicability of results from different projects. These efforts could lead to the establishment of a network of Living Labs, as envisioned in the European Commission's sustainable and smart mobility strategy [47].

3 Conclusions

Session outputs are based on all presentations, contributions and discussions. They include a compilation of key findings and lessons learned, as well as identified areas for future work and research. We hope that researchers, policymakers, and transportation planners can draw on these lessons and suggestions to develop engagement and planning strategies that effectively integrate public and stakeholder input to design AV systems that promote more equitable and sustainable outcomes based on a Quadruple Helix approach.

Summary of Key Findings and Lessons Learned:

- Communication and engagement projects have helped increase public support for AVs. Prior research has shown that effective communication (via knowledge raising and awareness campaigns); achieving familiarity and building confidence through direct experiences with AV technology (e.g., pilots and testing experiences in Japan and the US); and direct engagement with citizens and stakeholders (e.g., through cooperative planning exercises), contribute to increasing public support for AVs and allow for a joint understanding of AV implementation (i.e., vision, aims, objectives, challenges).
- Nevertheless, even amidst ongoing public outreach efforts, there is still evidence of widespread hesitation to buy or ride in an automated vehicle: This hesitation signals underlying concerns and values that need to be further explored and addressed.
- Developing intersectoral cooperation and promoting a joint understanding of automated mobility is a key requirement for successful AV development and deployment.

Beyond engaging with members of the public, transportation planners and AV developers should engage with public authorities who still lack knowledge about automated mobility and a clear strategy towards effective AV planning and implementation (EU experience). This cooperation is essential given the scale of the required ICT infrastructure.

- Mobility culture and mind-sets are changing globally. Through digitalisation and sharing-economy trends, there is potential for AVs to promote shifts in travel behaviour, especially among young people. Engagement processes are further supporting this trend by helping to promote widespread support (globally) for new forms of collective (and shared) automated mobility. When exploring AV preferences, both commuting and leisure journeys should be taken into account.

4 Next Steps

The following suggested next steps for future work and research emerged from this conference session:

- Develop a framework for coordinated research and innovation strategy among international authorities and programmes. Such a framework would help enhance public support by fostering cooperation and knowledge exchange.
- Researchers need to consider social, spatial, operational, and economic factors when developing research projects and testbeds aimed at developing AV services. Expanding the range of experts involved in project development could help to promote these broader environmental and social goals.
- Policies should foster new mobility services. Such services could help reduce the dependency on private owned cars (e.g., collective- and shared-services) and respond to user needs. Further, setting up international collaborative frameworks (e.g. CCAM) is essential and should be supported financially at all levels, particularly by the AV industry.
- Focus on narratives when communicating about AVs. An understandable and convincing narrative is key to reaching out to the public. Policymakers and developers should focus on using concrete examples and familiar concepts (e.g., AVs as a "horizontal elevator"). People can achieve nuanced understandings of AV technologies when the messaging is clear.
- Move from acceptance to co-creation. Planning processes must be interactive and grounded in mutual learning and cooperation based on relevant indicators (e.g. SUMINI) [48]. People need to be able to understand the full range of potential AV impacts [49]. When engaging with the public, it is important to shift the focus of the discussion from the technical aspects of AVs to the wider mobility goals that should promoted (e.g., equity, safety, accessibility, liveability).
- Experiment more with Living Labs. Living Labs provide opportunities to actively engage users through an open innovation framework and allow co-creation to advance towards a common vision for automated mobility. By enabling experimentation in a real-life context, Living Labs help to achieve a strategic and holistic perspective.

- Encourage citizen science projects that foster global co-creation. Such bottom-up approaches will inform policy makers at all levels that diverse communication, testing and implementation methods are needed to ensure strong public support for automation in the planning process.

Overall, this conference session emphasized the value of public engagement and outreach activities. Whilst much has been learned and gained from these activities, there is a lot more work remaining. To that end, more engagement and coordination activities are required at the local, national, and international level whilst shaping a more automated future founded on strong public support.

References

1. SAE: J3016B: taxonomy and definitions for terms related to driving automation systems for on-road motor vehicles, 15 June 2018. https://www.sae.org/standards/content/j3016_201806/. Accessed 19 Apr 2021
2. Williams, E., Das, V., Fisher, A.: Assessing the sustainability implications of autonomous vehicles: recommendations for research community practice. Sustainability **12**(5) (2020). Art. no. 5. https://doi.org/10.3390/su12051902
3. Masson-Delmotte, V., et al.: Global warming of 1.5 °C. An IPCC Special Report on the impacts of global warming of 1.5°C above pre-industrial levels and related global greenhouse gas emission pathways, in the context of strengthening the global response to the threat of climate change, sustainable development, and efforts to eradicate poverty (2018). Accessed 28 Oct 2021. https://www.ipcc.ch/sr15/
4. UNEP: 'Global Green New Deal'. Environmentally-Focused Investment Historic Opportunity for 21st Century Prosperity and Job Generation (2008)
5. UNFCC: Key aspects of the Paris Agreement, United Nations Climate Change. https://unfccc.int/process-and-meetings/the-paris-agreement/the-paris-agreement/key-aspects-of-the-paris-agreement. Accessed 28 Oct 2021
6. Milakis, D., Thomopoulos, N., Van Wee, B.: Policy Implications of Autonomous Vehicles. Academic Press (2020)
7. Thomopoulos, N., Givoni, M.: The autonomous car—a blessing or a curse for the future of low carbon mobility? An exploration of likely vs. desirable outcomes. Eur. J. Futures Res. **3**(1), 1–14 (2015). https://doi.org/10.1007/s40309-015-0071-z
8. Shiftan, Y., Polydoropoulou, A., Thomopoulos, N., Rappazzo, V.: Autonomous and connected transport - the user perspective, Special Issue. Sustain (2020)
9. Carayannis, E.G., Campbell, D.F.J.: Mode 3 knowledge production in Quadruple helix innovation systems. In: Carayannis, E.G., Campbell, D.F.J. (eds.) Mode 3 Knowledge Production in Quadruple Helix Innovation Systems: 21st-Century Democracy, Innovation, and Entrepreneurship for Development, pp. 1–63. Springer, New York (2012). https://doi.org/10.1007/978-1-4614-2062-0_1
10. Khoeini, S.: Autonomous vehicles: familiarity, awareness, and perceptions. Presented at the Automated Road Transportation Symposium 2021, 15 July 2021. https://www.tomnet-utc.org/leadership-webinarseries.html
11. Nair, G.S., Bhat, C.R.: Sharing the road with autonomous vehicles: perceived safety and regulatory preferences. Transp. Res. Part C Emerg. Technol. **122**, 102885 (2021). https://doi.org/10.1016/j.trc.2020.102885

12. Moody, J., Bailey, N., Zhao, J.: Public perceptions of autonomous vehicle safety: an international comparison. Saf. Sci. **121**, 634–650 (2020). https://doi.org/10.1016/j.ssci.2019.07.022
13. Eurobarometer: Expectations and concerns from a connected and automated mobility Eurobarometer survey, April 2020. Accessed 10 Sept 2021. https://europa.eu/eurobarometer/surveys/detail/2231
14. Kyriakidis, M., Sodnik, J., Stojmenova, K., Elvarsson, A.B., Pronello, C., Thomopoulos, N.: The role of human operators in safety perception of AV deployment—insights from a large european survey. Sustainability **12**(21) (2020). Art. no. 21. https://doi.org/10.3390/su12219166
15. Polydoropoulou, A., et al.: Who is willing to share their AV? Insights about gender differences among seven countries. Sustainability **13**(9) (2021). Art. no. 9. https://doi.org/10.3390/su13094769
16. Canitez, F., Thomopoulos, N., Cantafio, G.: WISE-ACT atlas: AV trials global overview. Presented at the WISE-ACT Workshop 1, Bratislava, Slovakia, 17 March 2018
17. Narayanan, S., Chaniotakis, E., Antoniou, C.: Chapter one - factors affecting traffic flow efficiency implications of connected and autonomous vehicles: a review and policy recommendations. In: Milakis, D., Thomopoulos, N., van Wee, B. (eds.) Advances in Transport Policy and Planning, vol. 5, pp. 1–50. Academic Press (2020). https://doi.org/10.1016/bs.atpp.2020.02.004
18. Thomopoulos, N., Cohen, S., Hopkins, D., Siegel, L., Kimber, S.: All work and no play? Autonomous vehicles and non-commuting journeys. Transp. Rev. **41**(4), 456–477 (2021). https://doi.org/10.1080/01441647.2020.1857460
19. Gkartzonikas, C., Gkritza, K.: What have we learned? A review of stated preference and choice studies on autonomous vehicles. Transp. Res. Part C Emerg. Technol. **98**, 323–337 (2019). https://doi.org/10.1016/j.trc.2018.12.003
20. Zegras, P.C., Butts, K., Cadena, A., Palencia, D.: Spatiotemporal dynamics in public transport personal security perceptions: digital evidence from Mexico City's periphery. ICT Transport, May 2015. Accessed 29 Oct 2021. https://www.elgaronline.com/view/edcoll/9781783471287/9781783471287.00014.xml
21. Etzioni, S., et al.: Modeling cross-national differences in automated vehicle acceptance. Sustainability **12**(22) (2020). Art. no. 22. https://doi.org/10.3390/su12229765
22. Moody, J., Middleton, S., Zhao, J.: Rider-to-rider discriminatory attitudes and ridesharing behavior. Transp. Res. Part F Traffic Psychol. Behav. **62**, 258–273 (2019). https://doi.org/10.1016/j.trf.2019.01.003
23. Krueger, R., Rashidi, T.H., Rose, J.M.: Preferences for shared autonomous vehicles. Transp. Res. Part C Emerg. Technol. **69**, 343–355 (2016). https://doi.org/10.1016/j.trc.2016.06.015
24. Kim, S.H., Circella, G., Mokhtarian, P.L.: Identifying latent mode-use propensity segments in an all-AV era. Transp. Res. Part Policy Pract. **130**, 192–207 (2019). https://doi.org/10.1016/j.tra.2019.09.015
25. Kassens-Noor, E., Kotval-Karamchandani, Z., Cai, M.: Willingness to ride and perceptions of autonomous public transit. Transp. Res. Part Policy Pract. **138**, 92–104 (2020 https://doi.org/10.1016/j.tra.2020.05.010
26. Dong, X., DiScenna, M., Guerra, E.: Transit user perceptions of driverless buses. Transportation **46**(1), 35–50 (2017). https://doi.org/10.1007/s11116-017-9786-y
27. Rahimi, A., Azimi, G., Asgari, H., Jin, X.: Adoption and willingness to pay for autonomous vehicles: attitudes and latent classes. Transp. Res. Part Transp. Environ. **89**, 102611 (2020). https://doi.org/10.1016/j.trd.2020.102611
28. Rahimi, A., Azimi, G., Jin, X.: Examining human attitudes toward shared mobility options and autonomous vehicles. Transp. Res. Part F Traffic Psychol. Behav. **72**, 133–154 (2020). https://doi.org/10.1016/j.trf.2020.05.001

29. WISE-ACT: Working Group 2 Thematic Report (2021)
30. Haboucha, C.J., Ishaq, R., Shiftan, Y.: User preferences regarding autonomous vehicles. Transp. Res. Part C Emerg. Technol. **78**, 37–49 (2017). https://doi.org/10.1016/j.trc.2017.01.010
31. Kaplan, L.: U.S. public forums on automated mobility. Presented at the Automated Road Transportation Symposium 2021, 15 July 2021
32. Russ, M.: Cities & citizens a dialogue on CCAM. Presented at the Automated Road Transportation Symposium 2021, 15 July 2021
33. Farooque, M., Lloyd, J., Tomblin, D., Quach, K.: Driverless cars and the public: right answers or wrong questions? Presented at the Automated Road Transportation Symposium, 15 July 2021
34. Yonei, Y.: Public acceptance of autonomous vehicle in Japan. Presented at the Automated Road Transportation Symposium 2021, 15 July 2021
35. Kaplan, L., Farooque, M., Sarewitz, D., Tomblin, D.: Designing participatory technology assessments: a reflexive method for advancing the public role in science policy decision-making. Technol. Forecast. Soc. Change **171**, 120974 (2021). https://doi.org/10.1016/j.techfore.2021.120974
36. Hilgarter, K., Granig, P.: Public perception of autonomous vehicles: a qualitative study based on interviews after riding an autonomous shuttle. Transp. Res. Part F Traffic Psychol. Behav. **72**, 226–243 (2020). https://doi.org/10.1016/j.trf.2020.05.012
37. Penmetsa, P., Adanu, E.K., Wood, D., Wang, T., Jones, S.L.: Perceptions and expectations of autonomous vehicles – a snapshot of vulnerable road user opinion. Technol. Forecast. Soc. Change **143**, 9–13 (2019). https://doi.org/10.1016/j.techfore.2019.02.010
38. Bennett, R., Vijaygopal, R., Kottasz, R.: Attitudes towards autonomous vehicles among people with physical disabilities. Transp. Res. Part Policy Pract. **127**, 1–17 (2019). https://doi.org/10.1016/j.tra.2019.07.002
39. Blumenthal, M.S., Fraade-Blanar, L., Best, R., Irwin, J.L.: Safe Enough: Approaches to Assessing Acceptable Safety for Automated Vehicles (2020). Accessed 29 Oct 2021. https://trid.trb.org/view/1751154
40. Kalra, N., Paddock, S.M.: Driving to safety: how many miles of driving would it take to demonstrate autonomous vehicle reliability? Transp. Res. Part Policy Pract. **94**, 182–193 (2016). https://doi.org/10.1016/j.tra.2016.09.010
41. Grosso, M.: Joint research centre focus groups with transport experts and non expert participants. Presented at the Automated Road Transportation Symposium 2021, 15 July 2021
42. Backhaus, W.: Automation-ready authorities fora with European cities. Presented at the Automated Road Transportation Symposium 2021, 15 July 2021
43. Uchimura, T.: Pilot test in Japan. Presented at the Automated Road Transportation Symposium 2021, 15 July 2021
44. ENoLL: What is ENoLL. European Network of Living Labs. http://enoll.org/about-us/. Accessed 8 Oct 2021
45. Martens, K.: Transport Justice: Designing Fair Transportation Systems. Routledge, New York (2016). https://doi.org/10.4324/9781315746852
46. Thomopoulos, N., Grant-Muller, S., Tight, M.R.: Incorporating equity considerations in transport infrastructure evaluation: current practice and a proposed methodology. Eval. Program Plann. **32**(4), 351–359 (2009). https://doi.org/10.1016/j.evalprogplan.2009.06.013
47. European Commission: COM(2020)789 - sustainable and smart mobility strategy - putting European transport on track for the future - Parlementaire monitor (2020). https://www.parlementairemonitor.nl/9353000/1/j9tvgajcor7dxyk_j9vvij5epmj1ey0/vlehreuol0wv. Accessed 8 Oct 2021

48. Thomopoulos, N., Grant-Muller, S.: Incorporating equity as part of the wider impacts in transport infrastructure assessment: an application of the SUMINI approach. Transportation **40**(2), 315–345 (2013). https://doi.org/10.1007/s11116-012-9418-5

Author Index

A
Assanis, Dimitris, 53
Auld, Joshua, 71

B
Backhaus, Wolfgang, 167
Baker, Denise, 71
Balachandran, Avinash, 85

C
Chen, Tiffany L., 85
Chou, Danielle, 71
Chow, Edward, 123
Ciuffo, Biagio, 167

F
Foss, Ann, 71
Franco, Daniel, 167

G
Goh, Jonathan Y. M., 85
Grosso, Monica, 167

H
Habibovic, Azra, 149
Han, Jihun, 53
Hooft, Floris, 149

I
Inoue, Hideo, 13

K
Kaplan, Leah, 167
Kimmel, Shawn, 123
Kortum, Katherine, 34
Kuznicki, Scott O., 140
Kuzumaki, Seigo, 13

L
Langer, Therese, 53
Lappin, Jane, 1
Leonard, John J., 85
Lidicker, Jeffrey, 53

M
Machek, Elizabeth, 71
Madland, Olav, 149
McGill, Stephen, 85
Mersky, Avi Chaim, 53
Meyer, Gereon, 42
Michelmann, Jakob, 42
Mihelic, Rick, 53

N
Nicols, George, 123

O
Orosz, Gabor, 53

R
Raposo, Maria Alonso, 167
Riggs, William, 71

G. Meyer and S. Beiker (Eds.): ARTSymposium 2021, LNMOB, pp. 185–186, 2023.
https://doi.org/10.1007/978-3-031-11112-9

Rosman, Guy, 85
Rupprecht, Siegfried, 167

S
Shladover, Steven E., 1
Shuman, Valerie, 1
Smith, Scott, 71
Sofos, Marina, 53
Stent, Simon, 85
Straub, Edward, 123
Sumwalt, Robert L., 23

T
Thomopoulos, Nikolas, 167
Turnbull, Katherine F., 110

V
Vreeswijk, Jaap, 149

Z
Zachäus, Carolin, 42